'22

Methods in Molecular Biology

Series Editor
John M. Walker
School of Life and Medical Sciences
University of Hertfordshire
Hatfield, Hertfordshire, UK

For further volumes:
http://www.springer.com/series/7651

For over 35 years, biological scientists have come to rely on the research protocols and methodologies in the critically acclaimed *Methods in Molecular Biology* series. The series was the first to introduce the step-by-step protocols approach that has become the standard in all biomedical protocol publishing. Each protocol is provided in readily-reproducible step-by-step fashion, opening with an introductory overview, a list of the materials and reagents needed to complete the experiment, and followed by a detailed procedure that is supported with a helpful notes section offering tips and tricks of the trade as well as troubleshooting advice. These hallmark features were introduced by series editor Dr. John Walker and constitute the key ingredient in each and every volume of the *Methods in Molecular Biology* series. Tested and trusted, comprehensive and reliable, all protocols from the series are indexed in PubMed.

Jasmonate in Plant Biology

Methods and Protocols

Second Edition

Edited by

Antony Champion and Laurent Laplaze

UMR DIADE - IRD, Montpellier, France

Humana Press

Editors
Antony Champion
UMR DIADE - IRD
Montpellier, France

Laurent Laplaze
UMR DIADE - IRD
Montpellier, France

ISSN 1064-3745 ISSN 1940-6029 (electronic)
Methods in Molecular Biology
ISBN 978-1-0716-0141-9 ISBN 978-1-0716-0142-6 (eBook)
https://doi.org/10.1007/978-1-0716-0142-6

Cover caption: Quantitative visualization of JA perception in vivo in the root tip of *Arabidopsis thaliana* using the Jas9-VENUS biosensor (Chapter 12). Copyright: Dr. Antoine LARRIEU (University of Leeds, UK).

This Humana imprint is published by the registered company Springer Science+Business Media, LLC, part of Springer Nature.
The registered company address is: 233 Spring Street, New York, NY 10013, U.S.A.

Preface

This second edition provides two reviews and fourteen up-to-date protocols related to the plant hormone jasmonate (JA). The chapters in this book are organized into three parts. JA is fast emerging as a central regulator of coordinated plant physiological and developmental responses to abiotic and biotic stresses. Parts I and II deal with the role of JA in plant physiology and development and in plant-biotic interactions, respectively. Recent years have also seen the emergence of new concepts and methods to quantify JA and analyze its signaling pathway. Part III provides contributions on methods related to jasmonate metabolism and signaling.

Montpellier, France

Antony Champion
Laurent Laplaze

Contents

Part III Metabolism and Signaling

Contributors

RICARDO AROCA • *Department of Soil Microbiology and Symbiotic Systems, Estación Experimental del Zaidín (CSIC), Granada, Spain*

IKA NURUL AZIZAH • *Department of Chemistry, Graduate School of Science, Tohoku University, Sendai, Japan*

VERONICA BASSO • *Laboratoire d'Excellence ARBRE, Centre INRA-Lorraine, UMR 1136 INRA-Université de Lorraine 'Interactions Arbres/Microorganismes', Champenoux, France*

STEPHANE BELLAFIORE • *IRD, Cirad, Univ Montpellier, IPME, Montpellier, France*

CATHERINE BELLINI • *Department of Plant Physiology, Umeå Plant Science Centre, Umeå University, Umeå, Sweden; Institut Jean-Pierre Bourgin, INRA, AgroParisTech, CNRS, Université Paris-Saclay, Versailles, France*

SUJATA BHARGAVA • *Botany Department, Savitribai Phule Pune University, Pune, India*

COLETTE BROEKGAARDEN • *Plant-Microbe Interactions, Department of Biology, Faculty of Science, Utrecht University, Utrecht, The Netherlands*

ANTONY CHAMPION • *UMR DIADE - IRD, Montpellier, France*

SIXUE CHEN • *Department of Biology, University of Florida, Gainesville, FL, USA; University of Florida Genetics Institute (UFGI), Gainesville, FL, USA; Proteomics and Mass Spectrometry, Interdisciplinary Center for Biotechnology Research (ICBR), University of Florida, Gainesville, FL, USA; Plant Molecular and Cellular Biology Program, University of Florida, Gainesville, FL, USA*

MAIMOUNA CISSOKO • *Laboratoire Mixte International Adaptation des Plantes et Microorganismes Associés aux Stress Environnementaux, Centre de Recherche de Bel-Air, Dakar, Senegal; Laboratoire Commun de Microbiologie, Centre de Recherche de Bel-Air, Institut de Recherche pour le Développement/Institut Sénégalais de Recherches Agricoles/Université Cheikh Anta Diop, Dakar, Senegal*

LISA DAVID • *Department of Biology, University of Florida, Gainesville, FL, USA; University of Florida Genetics Institute (UFGI), Gainesville, FL, USA*

CLAUDE Y. HAMANY DJANDE • *Department of Biochemistry, Centre for Plant Metabolomics Research, University of Johannesburg, Auckland Park, South Africa*

IAN A. DUBERY • *Department of Biochemistry, Centre for Plant Metabolomics Research, University of Johannesburg, Auckland Park, South Africa*

IVO FEUSSNER • *Department of Plant Biochemistry, Albrecht-von-Haller-Institute for Plant Sciences, University of Goettingen, Goettingen, Germany; Service Unit for Metabolomics and Lipidomics, Goettingen Center for Molecular Biosciences (GZMB), University of Goettingen, Goettingen, Germany; Department of Plant Biochemistry, Goettingen Center for Molecular Biosciences (GZMB), University of Goettingen, Goettingen, Germany*

DEBORA GASPERINI • *Department of Molecular Signal Processing, Leibniz Institute of Plant Biochemistry, Halle (Saale), Germany*

KRYSTELLE GRAY • *Laboratoire Mixte International Adaptation des Plantes et Microorganismes Associés aux Stress Environnementaux, Centre de Recherche de Bel-Air, Dakar, Senegal*

KENGO HAYASHI • *Department of Chemistry, Graduate School of Science, Tohoku University, Sendai, Japan*

CORNELIA HERRFURTH • *Department of Plant Biochemistry, Albrecht-von-Haller-Institute for Plant Sciences, University of Goettingen, Goettingen, Germany; Service Unit for Metabolomics and Lipidomics, Goettingen Center for Molecular Biosciences (GZMB), University of Goettingen, Goettingen, Germany*
VALÉRIE HOCHER • *Laboratoire des Symbioses Tropicales et Méditerranéennes, Institut de Recherche pour le Développement/INRA/CIRAD/Université Montpellier/SupagroM, Montpellier, France*
JIANING KANG • *Department of Biology, University of Florida, Gainesville, FL, USA; University of Florida Genetics Institute (UFGI), Gainesville, FL, USA; College of Life Science, Northeast Agricultural University, Harbin, Heilongjiang, China*
ABDELLAH LAKEHAL • *Department of Plant Physiology, Umeå Plant Science Centre, Umeå University, Umeå, Sweden*
LAURENT LAPLAZE • *UMR DIADE - IRD, Montpellier, France*
ANTOINE LARRIEU • *LS2 9JT, Leeds, UK*
MICHEL LEBRUN • *Institut de Recherche pour le Developpement (IRD), Cirad, Universite fe Montpellier, DIADE, Montpellier, France; Vietnam Academy of Science and Technology (VAST), LMI-RICE2, University of Science and Technology of Hanoi (USTH), Hanoi, Vietnam; IRD, Cirad, Univ Montpellier, LSTM, Montpellier, France*
YINGYAO LIU • *Temasek Life Sciences Laboratory, Singapore, Singapore*
NTAKADZENI E. MADALA • *Department of Biochemistry, Centre for Plant Metabolomics Research, University of Johannesburg, Auckland Park, South Africa*
HUONG TO THI MAI • *Vietnam Academy of Science and Technology (VAST), LMI-RICE2, University of Science and Technology of Hanoi (USTH), Hanoi, Vietnam*
KARSTEN MELCHER • *Laboratory of Structural Biology and Biochemistry, Van Andel Research Institute, Grand Rapids, MI, USA*
STEFAN MIELKE • *Department of Molecular Signal Processing, Leibniz Institute of Plant Biochemistry, Halle (Saale), Germany*
DANIEL MOUKOUANGA • *Institut de Recherche pour le Developpement (IRD), Cirad, Universite fe Montpellier, DIADE, Montpellier, France*
ASWATHY NAIR • *Botany Department, Savitribai Phule Pune University, Pune, India*
NAWEED I. NAQVI • *Temasek Life Sciences Laboratory, Singapore, Singapore*
MARIAMA NGOM • *Laboratoire Mixte International Adaptation des Plantes et Microorganismes Associés aux Stress Environnementaux, Centre de Recherche de Bel-Air, Dakar, Senegal; Laboratoire Commun de Microbiologie, Centre de Recherche de Bel-Air, Institut de Recherche pour le Développement/Institut Sénégalais de Recherches Agricoles/ Université Cheikh Anta Diop, Dakar, Senegal; Laboratoire Campus de Biotechnologies Végétales, Département de Biologie Végétale, Faculté des Sciences et Techniques, Université Cheikh Anta Diop, Dakar-Fann, Senegal*
TRANG HIEU NGUYEN • *Institut de Recherche pour le Développement (IRD), Cirad, Université de Montpellier, UMR DIADE, Montpellier, France*
RAJESH N. PATKAR • *Bharat Chattoo Genome Research Centre, Department of Microbiology and Biotechnology Centre, The Maharaja Sayajirao University of Baroda, Vadodara, India*
CORNÉ M. J. PIETERSE • *Plant-Microbe Interactions, Department of Biology, Faculty of Science, Utrecht University, Utrecht, The Netherlands*
ALOK RANJAN • *Department of Plant Physiology, Umeå Plant Science Centre, Umeå University, Umeå, Sweden*
MICHAEL RIEMANN • *Karlsruhe Institute of Technology (KIT), Botanical Institute, Karlsruhe, Germany*

BEATRIZ SÁNCHEZ-ROMERA • *Department of Soil Microbiology and Symbiotic Systems, Estación Experimental del Zaidín (CSIC), Granada, Spain*
MEREL STEENBERGEN • *Plant-Microbe Interactions, Department of Biology, Faculty of Science, Utrecht University, Utrecht, The Netherlands*
KAHO SUZUKI • *Department of Chemistry, Graduate School of Science, Tohoku University, Sendai, Japan*
SERGIO SVISTOONOFF • *Laboratoire Mixte International Adaptation des Plantes et Microorganismes Associés aux Stress Environnementaux, Centre de Recherche de Bel-Air, Dakar, Senegal; Laboratoire Commun de Microbiologie, Centre de Recherche de Bel-Air, Institut de Recherche pour le Développement/Institut Sénégalais de Recherches Agricoles/ Université Cheikh Anta Diop, Dakar, Senegal; Laboratoire des Symbioses Tropicales et Mé diterranéennes, Institut de Recherche pour le Développement/INRA/CIRAD/Université Montpellier/SupagroM, Montpellier, France*
MAME OURÈYE SY • *Laboratoire Mixte International Adaptation des Plantes et Microorganismes Associés aux Stress Environnementaux, Centre de Recherche de Bel-Air, Dakar, Senegal; Laboratoire Campus de Biotechnologies Végétales, Département de Biologie Végétale, Faculté des Sciences et Techniques, Université Cheikh Anta Diop, Dakar-Fann, Senegal*
YOUSUKE TAKAOKA • *Department of Chemistry, Graduate School of Science, Tohoku University, Sendai, Japan; Precursory Research for Embryonic Science and Technology (PREST), Japan Science and Technology Agency, Tokyo, Japan*
H. V. THULASIRAM • *Chemical Biology Unit, Division of Organic Chemistry, CSIR-National Chemical Laboratory, Pune, India*
MINORU UEDA • *Department of Chemistry, Graduate School of Science, Tohoku University, Sendai, Japan; Department of Molecular and Chemical Life Sciences, Graduate School of Life Sciences, Tohoku University, Sendai, Japan*
CLAIRE VENEAULT-FOURREY • *Laboratoire d'Excellence ARBRE, Centre INRA-Lorraine, UMR 1136 INRA-Université de Lorraine 'Interactions Arbres/Microorganismes', Champenoux, France*
SASKIA C. M. VAN WEES • *Plant-Microbe Interactions, Department of Biology, Faculty of Science, Utrecht University, Utrecht, The Netherlands*
SHENG YANG HE • *Department of Energy Plant Research Laboratory, East Lansing, MI, USA*
FENG ZHANG • *College of Plant Protection, Nanjing Agricultural University, Nanjing, China*

Part I

Physiology and Development

Chapter 1

Multiple Roles of Jasmonates in Shaping Rhizotaxis: Emerging Integrators

Abdellah Lakehal, Alok Ranjan, and Catherine Bellini

Abstract

The root system and its architecture enormously contribute to plant survival and adaptation to the environment. Depending on the intrinsic genetic information and the surrounding rhizosphere, plants develop a highly plastic root system, which is a critical determinant for survival. Plant root system, which includes primary root (PR), lateral roots (LR) and adventitious roots (AR), is shaped by tightly controlled developmental programs. Phytohormones are the main signaling components that orchestrate and coordinate the genetic information and the external stimuli to shape the root system patterning or rhizotaxis. Besides their role in plant stress responses and defense against herbivory and pathogen attacks, jasmonic acid and its derivatives, including the receptor-active conjugate jasmonoyl-L-isoleucine (JA-Ile), emerge as potential regulators of rhizotaxis. In this chapter, we summarize and discuss the recent progress achieved during the recent years to understand the JA-mediated genetic and molecular networks guiding PR, LR, and AR initiation. We highlight the role of JAs as critical integrators in shaping rhizotaxis.

Key words Jasmonates, Rhizotaxis, Primary root, Lateral roots, Adventitious roots, Organogenesis

1 Introduction

Jasmonic acid (JA) and its derivatives, collectively called jasmonates (JAs), are lipid-derived phytohormones. Historically, one of the pioneer experiments in JA research has been performed as early as 1899 by Hesse and Muller [1] who extracted jasmone compound from jasmine flower (*Jasminium grandiflorum*). Later, Demole et al. [2] extracted and characterized the structure of the methyl ester of JA (MeJA) which was detected as an odorant of *Jasminium grandiflorum* flowers. Since then, mechanisms involved in JA biosynthesis, metabolism, transport, and signaling have been extensively studied, which contributed to improving our knowledge about its mode of action in numerous stress responses and more recently in developmental processes [3–5]

In the current model of JA perception and signaling, the activity of JA-responsive transcription regulators, such as the *bHLH*

Antony Champion and Laurent Laplaze (eds.), *Jasmonate in Plant Biology: Methods and Protocols*, Methods in Molecular Biology, vol. 2085, https://doi.org/10.1007/978-1-0716-0142-6_1, © Springer Science+Business Media, LLC, part of Springer Nature 2020

transcription factors *MYC2/JASMONATE-INSENSITIVE1* (*MYC2/JIN1*), *MYC3* or *MYC4*, is repressed by the JASMONATE ZIM DOMAIN (JAZ) proteins in the cells containing a low level of the bioactive form jasmonoyl-L-isoleucine (JA-Ile) [6, 7]. JAZ proteins recruit the general corepressor TOPLESS (TPL) and TPL-related proteins either directly or indirectly through interaction with the coadaptor protein NOVEL INTERACTOR OF JAZ (NINJA) to form a repressor complex [8, 9]. When JA-Ile concentration increases, it acts as a molecular glue between the F-box protein CORONATINE INSENSITIVE1 (COI1) and JAZ proteins to form the JA coreceptor complex [10]. This leads to ubiquitylation and subsequently degradation of the JAZ proteins via the 26S proteasome pathway, releasing the transcriptional activity of JA-responsive transcription factors [6, 7, 10].

Besides their prominent role in plant defense, JAs were shown to affect several aspects of plant development, such as embryo and/or seed development [11], hypocotyl elongation [12], flower organ development [13–15], flowering time [16], leaf growth and senescence [17, 18], primary root growth [19, 20], lateral root formation [21], and adventitious rooting [22, 23]. In this chapter, unless specified, we will focus on *Arabidopsis thaliana* research and we will highlight and discuss recent advances made in understanding the role of JAs in the control of root development. We also discuss the fact that JAs might be potential signaling hormones in shaping root architecture in natural conditions and in response to environmental cues.

2 Multiple Roles of JA Signaling in Primary Root Patterning and Elongation

The primary root (PR) is established during early embryogenesis. At very early stages of zygote divisions, a single cell called hypophysis, is recruited to form the PR founder cells. The hypophysis divides asymmetrically to establish PR meristematic cells, which includes the quiescent center (QC) and columella stem cells (CSCs) (reviewed in [24]). Auxin and cytokinins (CKs) are the main growth factors involved in the specification of the hypophysis and the establishment of the PR meristem during embryogenesis. They have an antagonist role and recently other hormones, such as gibberellins and brassinosteroids were suggested to crosstalk with auxin and/or CKs during the specification of the hypophysis and establishment of the root meristem (reviewed in [24]). After germination, the stem cell niche located at the root tip gives rise to all the cells that will form the PR through distinct and successive developmental phases: the meristematic zone (MZ) in which cells undergo several rounds of mitosis to form a pool of cells that will later elongate and differentiate; the elongation zone (EZ) in which the cells lose their ability to divide, but will elongate; and the differentiation zone (DZ) where the cells acquire their specificities and functions [24]. These coordinated events contribute to the primary root patterning and elongation, which are controlled by the crosstalk of several hormones including JAs. Growth inhibition was one of the first physiological effects of JAs that has been studied [25] and more recent work showed that JA-induced growth inhibition involves phase-specific disruption of cell cycle progression. Application of JAs on synchronized tobacco (*Nicotiana tabacum*) BY-2 cells induced an arrest of cell division in both G1- and G2-phases [26]. Whole-genome expression profiling of Arabidopsis cell cultures indicated that JAs repressed the activation of M-phase and consequently arrested the cell cycle in the G2-phase [27]. In line with this, Chen et al. [28] showed that exogenously applied JAs decreased the final cell length in the DZ and reduced the size of the EZ as well as the size of the MZ. This was due to a reduction both of cell number and of cell length, indicating that JAs reduce the two basic cellular processes affecting primary root growth, cell elongation, and division. Besides its action in inhibiting cell division and expansion in the PR tip, JAs also destabilize the PR stem cell niche by enhancing divisions in the QC and differentiation of columella stem cells [28]. In the root, QC identity and stem cell stability are controlled mainly by two auxin-induced *APETALA2 (AP2)* class of transcription factors *PLETHORA1* (*PLT1*) and *PLT2* that act redundantly [29]. The double mutant *plt1plt2* failed to establish and maintain a proper QC identity, which caused strong defects in PR patterning and growth [29]. Chen et al. [28] observed that the PR phenotype of the double mutant *plt1plt2* resembled that of JA-treated seedlings. Indeed, exogenously applied JAs induced cell proliferation in the QC and the formation of additional columella cell layers. The expression of *PLT1* and *PLT2* was shown to be repressed by JAs in a *COI1*- and *MYC2*-dependent manner. It was confirmed that the MYC2 protein bound directly the *PLT1* and *PLT2* promoter regions to repress their expression (Fig. 1 and Table 1) [28]. Thus, it is likely that *MYC2*-mediated JA repression of *PLT1* and *PLT2* expression contributes to integrate JAs and auxin signaling pathways to modulate the root stem cell niche.

The evidence for the role of JAs in the control of PR development came primarily from the characterization of *Arabidopsis* mutants affected in JA perception or signaling. Many components in JA perception and signaling have been identified in genetic screens looking for either insensitivity or hypersensitivity to exogenous JA treatment. Although most of these mutants show a PR defect in response to exogenous JA treatment, the specific contribution of the corresponding genes to PR development is not clear yet. Indeed, genetic evidence for the role of single JA perception or signaling components is hampered by high functional redundancy among gene families [30]. For example, single loss-of-function

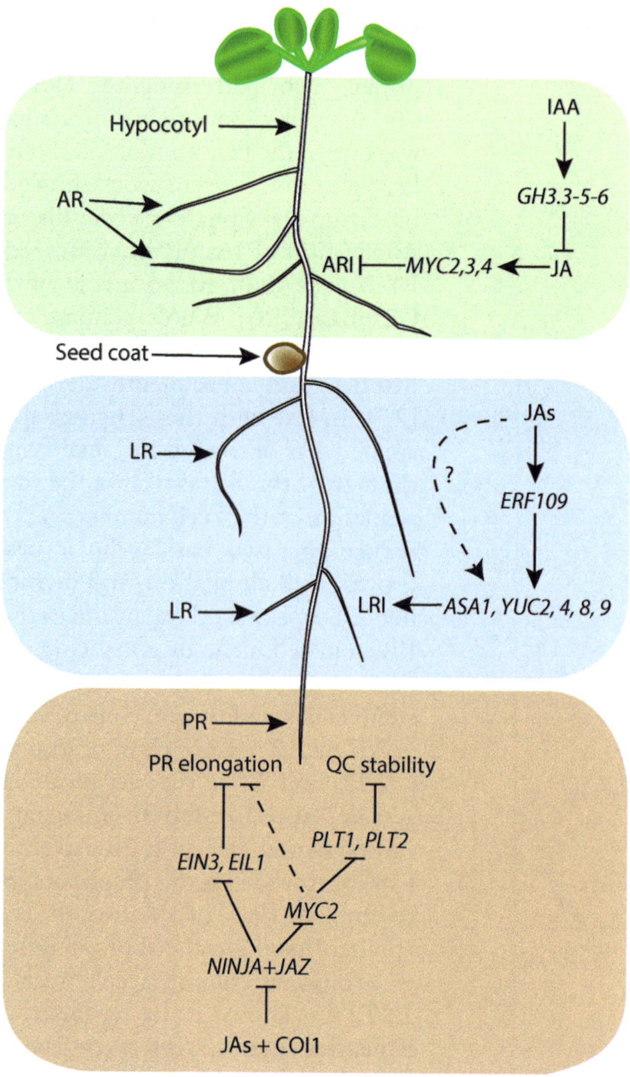

Fig. 1 Integrative model for the role of JAs in rhizotaxis. Auxin-inducible *GH3.3-5-6* inactivates the JA pool in the elongated hypocotyl and subsequently diminishes the availability of JA-Ile, which is the bioactive form that triggers JA signaling. *MYC2-3-4*-mediated JA signaling additively inhibits ARI with a mechanism yet to be identified. JAs promote LRI by inducing the expression of *ERF109*, which directly binds to the promoters of key genes in L-Trp-derived IAA biosynthesis, *ASA1* and *YUC2*. JAs also activate the transcription of other IAA biosynthesis genes in the same pathway, such as *YUC4*, *YUC8*, and *YUC9*. JAs inhibit PR elongation and destabilize the stem cell niche by interfering with QC maintenance. MYC2 directly binds to *PLT1* and *PLT2* promoters to repress their transcription. *PLT1* and *PTL2* are required for proper QC identity and stability. JAs derepress the transcriptional activity of EIN3 and EIL1 by degrading JAZ repressors. *EIN3* and *EIL1* redundantly inhibit PR elongation

Table 1
List of genes (transcriptionally) regulated by JA involved in rhizotaxis

Gene	Description	Regulation by JAs	PR	LR	AR	References
JA-related						
COI1	JA receptor	Up	–	+	–	[22, 28, 62]
MYC2	A master TF in JA signaling	Up	–	+	–	[22, 28, 35, 63]
JAZ family	Components of JA-signaling repressor complex	Up	+	?	?	[33]
NINJA	Mediates *TPL-JAZ* interactions to repress JA signaling	Up	+	?	?	[34, 35]
GH3.11	Catalyzes the conversion of JA into JA-Ile	Up	–	?	–	[22]
JAT1	Cellular JA and JA-Ile transporter		–	?	?	[39]
IAA-related						
GH3,5-6	Catalyze the conversion of JA into JA-Trp, JA-Met and JA-Asp	Up	?	?	+	[22, 80, 81, 63]
ARF17	Auxin response factor	Up	?	?	–	[22, 80, 81, 84]
PLT1,PLT2	AP2-domain TF, key player in QC stability	Down	+	?	?	[28, 29]
LBD26	LOB-domain TF, control LRI	Down	?	+	?	[66]
PUCHI	ERF TF, control LRI	Down	?	+	?	[66]
PIN1, PIN2	IAA efflux carries	Up	+	+	+	[59, 92, 93]
ET-related						
ERF109	Controls the expression of *ASA1* and *YUC2*	Up	–	+	?	[47]
Cell cycle						
CYCB1;1	Mediates cell division at G2/M phase	Down	+	+	+	[28, 84]

+ Positive regulator, – Negative regulator, ? Not investigated yet mutants in any gene from the *JAZ* family did not show any obvious PR phenotype either in the presence or in the absence of exogenous JAs [31]. The Arabidopsis genome contains 13 *JAZ* members and only multiple mutants, such as the quintuple mutant (*jazQ*), which harbor loss-of-function mutations in five members of the *JAZ* family (*JAZ1, 3, 4, 9,* and *10*) exhibited a slightly shorter root compared to the wild type [32]. Mutating ten out of 13 members of the *JAZ* family (*JAZ1-7, -9, -10, -13*) not only further reduced the PR length, but also increased the sensitivity to exogenously applied JAs [33] confirming the high functional redundancy among *JAZ* genes.

In a genetic screen for constitutive activity of the JA-inducible *pJAZ10:GUSplus* marker, Acosta et al. [34] identified three loss-of-function alleles in the NINJA coadaptor protein. Interestingly, *ninja* loss-of-function mutants exhibited defect in cortical cell elongation, which were 20% shorter than in wild type, in the DZ of the PR and consequently displayed a shorter PR compared to that of the wild type [34]. These results confirmed that *NINJA* is a key component of the JA signaling repressor complex and its removal allowed JA-dependent signaling to inhibit cell expansion in the DZ, but not cell division in the MZ. Strikingly, when the *ninja* mutation was combined with the dosage-dependent gain-of-function mutant *myc2-322B*, the PR was strongly inhibited and it resembled JA-treated roots [35]. The shorter root in double mutant *ninja myc2-322B* was due to defects in cell expansion and cell division. In line with this, *ninja myc2-322B* was extremely hypersensitive to exogenous MeJA treatment [35]. It is noteworthy that, to our knowledge, double mutant *ninja myc2-322B* is the first JA signaling mutant with such a strong developmental phenotype. Considering that *NINJA* might also be involved in other signaling pathways, double mutant *ninja myc2-322B* would definitely be a valuable genetic tool to investigate the role of JA signaling in PR development.

Despite their distinct but slightly overlapping expression pattern in the root tip [35], the transcription factors *MYC3* and *MYC4* act additively with *MYC2* to mediate JA responsiveness in terms of PR inhibition [36]. In contrast to loss-of-function mutant *myc2*, loss-of-function mutants *myc3* and *myc4* and even double mutant *myc3myc4* were fully sensitive to JA treatment. Nevertheless, the double mutant *myc3myc4* enhanced the insensitivity of *myc2* in response to exogenous JA treatment [36]. These data indicate that *MYC2* is the master regulator of JA-induced PR inhibition and the absence of the phenotype in *myc3myc4* background is likely due to a compensatory effect by a possible upregulation of *MYC2* in the double mutant *myc3myc4*.

In conclusion, *MYC2*, as master regulator and hub of JA signaling, mediates distinct roles to control PR growth and patterning (Fig. 1). Besides its action in inhibiting cell division and expansion in the DZ of the PR tip, MYC2 also destabilizes the PR stem cell niche activity by enhancing QC cell divisions and columella stem cell differentiation [28]. Nevertheless, the molecular and cellular bases of these complex and distinct roles of MYC2-mediated JA signaling during PR development are not well understood yet. More surprisingly, the mechanistic bases of the role of JAs in the inhibition of cell expansion and elongation in the PR are also missing. Cell expansion and elongation are the result of multiple mechanisms including turgor pressure, endoreduplication, microtubules' organization and orientation as well as cell wall modifications. JAs have been shown to modulate several of these mechanisms. For example, JAs repress endoreduplication during leaf expansion [18], they affect the expression of several cell wall-modifying enzymes and might therefore affect cell wall rigidity or flexibility [27, 37]. Indeed, it would be interesting to check if JAs affect cell expansion, directly or indirectly via auxin or whether they interfere with the expression and/or the function of cell wall-modifying enzymes in the EZ of the PR to inhibit cell expansion.

3 Potential Role of JA Transport During PR Elongation

The role of JA transport in plant development is poorly studied and only few reports highlighted its importance in PR growth. For example, Gasperini et al. [38] demonstrated that JA signaling induced in the shoot by mechanical wounding was able to reach the root where it repressed cell division and elongation in the MZ causing a reduction of PR length. In contrast, the loss-of-function mutant *aos*, altered in the JA biosynthesis enzyme ALLENE OXIDE SYNTHASE, did not show any PR length defect after wounding [35, 38]. These data suggest that wound-induced PR inhibition is dependent on AOS-mediated JA biosynthesis. It can be hypothesized that JA and /or JA-Ile derived from the wounding site translocate from shoot to root to inhibit PR elongation. In line with this, a cellular JA transporter *JASMONATE TRANSPORTER1/ATP-BINDING CASSETTE G16 (JAT1/ABCG16)* was recently identified and implicated in the control of PR growth and other JA-mediated processes. The loss-of-function mutant *jat1-1* exhibited bigger root cell area and reduced sensitivity to exogenous application of JAs, relatively to the wild type, indicating the importance of JA transport in PR growth [39]. This was the first report on JA transport, thus it will be exciting in the future to know more about cell-to-cell JA transport and its physiological significance in mediating rapid JA responses to environmental cues as well as their contribution in plant growth and development.

4 JAs Integrate Multiple Hormonal Pathways to Control PR Development

PR development is controlled by complex hormonal circuits and feedback loops. Auxin, CKs and ethylene (ET) are the most studied ones and their crosstalk during this process is partly unveiled [24]. Despite their importance in PR development, the interaction of JAs with other phytohormones is poorly understood. Like most hormones, JAs act in a dose-dependent manner. It was shown that JAs interfered with polar auxin transport by reducing PIN1 and PIN2 protein level at the plasma membrane, but depending on the concentration the mechanisms involved are different [40]. At high concentration JAs reduced PIN2 endocytosis in a COI1 and AUXIN RESISTANT 1 (AXR1)-dependent manner [40]. *AXR1* encodes a subunit of the RUB1-activating enzyme, which is a key component of several hormonal signaling pathways including JAs [41]. In contrast, the inhibitory effect of low concentration of JAs on PIN2 endocytosis was reduced in the mutant *asa1* and in the quadruple mutant *tir1afb1,2,3* [40] confirming that at low JA concentration, the effect on the intracellular trafficking of the PIN proteins requires WEAK ETHYLENE INSENSITIVE2/ ANTHRANILATE SYNTHASE alpha1 (WEI2/ASA1)-dependent auxin biosynthesis and the ubiquitin-protein ligase TRANSPORT INHIBITOR RESPONSE1/AUXIN SIGNALING F-BOX (TIR1/AFB)-dependent auxin signaling. *TIR1/ AFBs* encode for nuclear auxin receptors [42, 43] and *WEI2/ ASA1* encodes an enzyme catalyzing the rate limiting step in tryptophan-dependent indole-3-acetic acid (IAA) biosynthesis [44]

ET is another important hormone controlling PR elongation. Zhu et al. [45] reported that *ETHYLENE INSENSITIVE 3* (*EIN3*) and its closest paralog *EIN3-LIKE1* (*EIL1*) integrate ET and JA signaling to control PR growth. The double mutant *ein3eil1* was partially insensitive to JA-induced PR growth inhibition, whereas *EIN3* or *EIL1* overexpressing lines were hypersensitive to JA in term of PR length [45]. At the molecular level, JAZ repressors physically interacted with ET-stabilized EIN3 to repress its transcriptional activity, while JA treatment mediates JAZ degradation and thereby releases EIN3 transcriptional activity (Fig. 1) [45]. This suggests that JAs synergistically interact with ET to control PR elongation. In line with this, *ETHYLENE RESPONSE FACTOR1* (*ERF1*), which is induced both by exogenous ET or JA treatments in a *EIN3* and *COI1*-dependent manner, plays a crucial role in PR elongation [46]. Downregulated *RNAi-erf1* lines exhibited slightly longer PR, whereas *ERF1* overexpressing lines exhibited shorter PR compared to their wild-type counterparts [46]. Mechanistically, *ERF1* enhances IAA biosynthesis to inhibit PR elongation by directly controlling the expression of *WEI2/ ASA1* specifically in the PR tip [46]. Similarly, *ERF109*, which is highly and rapidly induced by exogenously applied MeJA, plays an important role in mediating crosstalk between JAs and auxin [47].

Another prime example for the role of JAs in modulating complex signaling circuits is the interaction with brassinosteroids (BRs). BRs mediate myriad of developmental processes including PR and LR development [48]. A detailed investigation of BRs biosynthesis and signaling mutants revealed that BRs mediate several hormonal inputs to provide a correct information for PR development. BRs control both cell elongation and stem cell niche stability during PR growth and in interaction with other phytohormones including JAs [48]. In a genetic screen looking for suppressor of *coi1-2* insensitivity to MeJA-induced PR inhibition, Ren et al. [49] identified the *partially suppressing coi1* (*psc1*) mutant, which suppressed partially the MeJA insensitivity phenotype of *coi1-2*. *psc1* is a recessive mutant in the *DWARF4* (*DWF4*) gene, which encodes an enzyme catalyzing a rate-limiting step in BR biosynthesis [50]. Further experiments showed that *psc1/dwf4* was hypersensitive to MeJA, whereas the dominant line (*gulliver3-D/dwarf4-5D*) was partially insensitive to MeJA in terms of PR inhibition [49, 51]. In addition, exogenous application of BR restored the insensitivity to MeJA-induced PR inhibition of the double mutant *coi1psc1* and MeJA treatment downregulated the expression of *DWF4* in a COI1-dependent manner, confirming that *DWR4* acts downstream of *COI1* [49]. Although endogenous BR content needs to be quantified upon JA treatment, this suggests that COI1-dependent JA signaling negatively regulates BR biosynthesis to inhibit PR elongation.

Interaction between JAs and CKs remains poorly understood, these hormonal pathways can interplay during plant growth and development as well as plant defense against biotic and abiotic stress. JAs seem to synergistically or antagonistically interact with CKs, depending on the species and the context. For example, physiological studies showed that MeJA promotes CK accumulation in wheat seedlings under salinity by repressing the *CYTOKININ OXIDASE/DEHYDROGENASE*-mediated degradation [52]. Jang et al. [53] showed that treatment with JAs impaired the CK response in Arabidopsis root through a mechanism yet to be identified. COI1-dependent JA signaling induced extra xylem formation in the Arabidopsis root and treatment with the synthetic CK 6-Benzyl-aminopurine abolished this effect. This indicates that JAs and CKs interact antagonistically to control root xylem formation [53].

5 JAs Mediate Environmental Cues to Control PR Growth

Despite the little information about the role of JAs in guiding the developmental plasticity of the PR in response to diverse environmental conditions, it would not be surprising if JAs would appear as critical integrators of environmental stimuli and PR growth and development. For example, Yang et al. [54] proposed that aluminum inhibition of root growth is mediated by JA signaling in a COI1-dependent manner, indicating that JAs are an important mediator of environmental stresses [54]. In conclusion, the above results provide strong evidence for the role of JAs as integrators of complex hormonal and environmental signaling pathways during PR development.

6 Role of JAs in Regulating LR Development

Plant adaptation and survival depend on root system plasticity, which is, in part, controlled by three integrated and synchronized developmental LR development processes: LR initiation (LRI), distribution (spacing), and elongation. LRs play a critical role in water and nutrient uptake by exploring their surrounding environment. In the model plant *Arabidopsis*, lateral root development starts with priming of the pericycle cells at the root basal meristem. Only specific subset of pericycle cells neighboring to the protoxylem tissues receives a discrete signal, which makes them primed to form LR founder cells. Some LR founder cells can then undergo successive cell divisions to form LR primordia. LR formation in Arabidopsis has been extensively studied and thoroughly reviewed [24, 55–57].

It is now evident that JAs play an important role in LR formation [58, 59]. Physiological studies showed that low concentrations (micromolar range) of JAs promoted LR formation in Arabidopsis and in rice (*Oryza sativa*) [47, 59–61]. In line with this, the Arabidopsis loss-of-function mutant *coi1-1* was shown to be partially insensitive to JA-induced lateral rooting compared to the wild type indicating that the ability of JAs to promote LR development is in part mediated by the COI1 receptor [59, 62]. The mutant *coi1-1* showed a decrease in stage I LR primordia suggesting that COI1 is required for early event of pericycle divisions during LRI [62]. In addition, it was shown that *MYC2*, which exhibited a strong root-specific expression acts as a positive regulator of LR development [63]. In this process, *MYC2* acts additively with *SUPPRESSOR OF PHYTOCHROME A-105 1* (*SPA1*), a regulator of phytochrome signaling, which is also required for JA-mediated LR induction [64, 65].

7 JAs Act on IAA Biosynthesis and Signaling to Control LR Architecture

A genetic screen for suppressors of JA-mediated LR induction revealed that JAs act through ASA1-mediated IAA biosynthesis to promote LR formation [59]. It has also been shown that COI1-

dependent JA signaling mediates the expression of *ASA1*, *YUCCA2 (YUC2)*, *YUC8*, and *YUC9* (Fig. 1) that encode enzymes involved in auxin biosynthesis [47, 59, 61]. Arabidopsis genome contains 11 *YUC* genes, which catalyze the conversion of indole-3-pyruvic acid (IPyA) into IAA. Therefore, JAs promote the L-Tryptophan-dependent IAA biosynthesis by inducing the expression of the key genes involved in this pathway. JAs strongly induced IAA response, as marked by the auxin inducible construct *pIAA12: GUS*, in the root basal meristem of wild type plants, but it strongly suppressed it in the root meristem of the loss-of-function mutant *asa1* [59], implying that JA signaling requires functional L-tryptophan-dependent IAA biosynthesis to promote LRI. The mechanistic insight into how JAs suppress IAA signaling in mutant *asa1* is still enigmatic and would require additional investigation.

Recently, *ERF109* has been proposed as a missing link between JAs and IAA biosynthesis (Fig. 1) [47]. MeJA rapidly induces the expression of *ERF109* in a *COI1*-dependent manner, and the ERF109 transcription factor binds directly to the GCC-box containing the regulatory element of the *ASA1* and *YUC2* promoters to regulate their transcription. *ASA1* and *YUC2* were transcriptionally induced in *35S:ERF109* expressing lines and repressed in the loss-of-function mutant *erf109* [47]. In line with this, the transgenic lines overexpressing *ERF109* exhibited strong pleiotropic phenotypes reminiscent of IAA overproducing mutants including an increased number of LR, whereas loss-of-function mutant *erf109* exhibited a slightly reduced LR number compared to the wild type [47]. Interestingly, the introgression of the *yuc2* loss-of-function mutation into the *35S:ERF109* background partially reversed the LR phenotype, confirming that JA-inducible *ERF109* controls YUC2-mediated IAA biosynthesis (Fig. 1 and Table 1) [47].

It was also shown that treatments with high concentration (20 or 50 μM) of (−) –JA or (+) –JA strongly reduced LR number in the wild type and in *coi1* knock-out mutants, indicating that (−) –JA or (+) –JA inhibit LR formation in a COI1-independent manner [66]. This reduction was also independent of ASA1-mediated IAA biosynthesis because both isomers could still induce the expression of *ASA1* and the subsequent IAA accumulation. In addition, concentration as high as 50 μM of (−) –JA-L-Ile or (+) –JA-L-Ile did not impair the LR number, confirming that (−) –JA or (+) –JA suppresses LR development independently of the JA-L-Ile-mediated COI1-JAZ signaling machinery [66]. In an attempt to explain the negative effect of JAs on the number of LRs, Ishimaru et al. [66] showed that both (−) –JA or (+) –JA not only strongly diminished auxin response and signaling, but also inhibited the expression of key genes involved in LRI, such as *GATA23*, *LATERAL ORGAN BOUNDARIES-DOMAIN 29*, and *PUCHI* (Table 1). The expression of *LBD29* and *PUCHI* is regulated by

AUXIN RESPONSE FACTOR genes *ARF7* and *ARF19* whose transcriptional activity is controlled by AUXIN/INDOLE-3-ACETIC ACID (Aux/IAA) repressors. This suggests that JAs likely inhibit auxin response by stabilizing Aux/IAA repressors, which are major components of the canonical auxin sensing machinery [66]. In addition, the COI1-independent negative effect of JAs on LRI was maintained in the double mutant *tir1afb2*, but not in the *afb5* mutant, which shows resistance to exogenously applied JAs compared to wild type [66]. Therefore, it seems that AFB5 might have a major role compared to TIR1, leading to a reduced Aux/IAA breakdown and inhibition of LR by JAs. Although the concentrations used are high and might not reflect *in planta* physiological concentrations, shedding light on the molecular and genetic bases of how JAs suppress auxin response in the context of LRI will be exciting and will bring new thoughts about JAs-IAA crosstalk, besides the established linear interaction, that is, JA induction of IAA biosynthesis.

8 Can JAs Be Central Integrators of the Mechanical Stimuli that Trigger LR Formation in Natural Condition?

In natural conditions, roots forage through the soil and subsequently encounter several environmental and mechanical constraints, which trigger complex signaling pathways involved in shaping root architecture [67]. Hence, it would not be surprising if stress-associated phytohormones play a crucial role during LRI in natural conditions. It was proposed that mechanical cues control LRI and their spacing independently of auxin transport and signaling, but the mechanism driving this process is still largely unclear [68, 69]. Auxin signaling double mutant *arf7arf19* or *solitary root-1* (*slr-1/iaa14*) did not produce any or very few LR when grown in vitro on agar plates placed vertically, whereas both formed LR when grown in soil or upon mechanical stimulation [68, 69]. The double mutant *arf7arf19* harbors loss-of-function mutations in *ARF7* and *ARF19*, whereas *slr-1/iaa14* harbors a mutation in domain II of the *IAA14* gene, which makes it resistant to auxin-dependent proteasome degradation. In addition, LR formation upon transient mechanical bending did not require the auxin receptor *TIR1* [68]. Based on these data, we can speculate that while the root tip penetrates through soil it constantly encounters mechanical and environmental stimuli which, independently of the *TIR1-IAA14-ARF7/ARF19* signaling module, control LRI. How the mechanical signals are perceived, transduced, and dynamically translated into synchronized developmental output is not known yet. Interestingly, *coi1-1* showed a strong resistance to gravistimulation or bending-induced LR formation [62] suggesting that COI1 receptor is an essential mediator of mechanical signals involved in LR patterning. Furthermore, touch-induced morphogenesis or thigmomorphogenesis in the shoot induced JA accumulation and required a functional COI1-dependent signaling for its action [70], providing evidence for the role of JAs in mediating mechanical-induced signals. Recently, a mechanosensing receptor kinase *FERONIA* (*FER*) has been shown to negatively regulate JA signaling by phosphorylating and destabilizing MYC2 transcription factor and thus contributing to plant immunity [71] and likely other developmental programs. Interestingly, the loss-of-function mutant *fer-4* exhibited an increased number of LR [72], but the molecular bases mediating or establishing this phenotype are not known yet. We speculate that in natural conditions, *FER* and/or other mechanosensing receptors [73] might perceive diverse environmental inputs and thereby change the JA biosynthesis and/or signaling balance at specific cell types. The perceived extrinsic input, along with the intrinsic genetic competence, triggers specific transcriptional reprogramming leading to a fine-tuned LR organogenesis output. Fast-developing phenomic techniques in soil, such as high-resolution X-ray scanners [74] and the development of JA perception biosensor, such as the Jas9-VENUS that allows the quantification of dynamic changes in JA distribution in response to stress with high spatiotemporal sensitivity [75] should definitely help study the significance and role of JAs in mediating and translating the environmental cues into strict developmental output to guide LR organogenesis in natural conditions or under physical constraints.

9 Role of JAs in Regulating Adventitious Rooting

Adventitious rooting is a highly plastic developmental process. Although adventitious root initiation (ARI) and LRI share a part of their genetic network, AR formation is also guided by specific and unique ones [57, 76]. A continuous treatment with low concentration of 12-oxo-phytodienoic acid (OPDA), JA, or MeJA inhibited AR formation in Arabidopsis etiolated hypocotyl, Petunia leafy cuttings, and *Bupleurum kaoi* indicating a negative role of JAs in ARI [22, 77, 78]. In line with these observations, genetic approaches using Arabidopsis etiolated hypocotyl as a model system strongly supported the inhibitory role of JAs. The loss-of-function mutant *coi1-16* produced more AR compared to the wild type, the single loss-of-function *myc2*, and the triple mutant *myc2myc3myc4* produced more AR, whereas the overexpressing *35S:MYC2* line produced fewer AR compared to the wild type, indicating that JAs inhibit ARI through the canonical COI1-MYC2-dependent pathway in Arabidopsis hypocotyl (Fig. 1 and Table 1) [22]. Despite its clear negative effects on AR initiation in Arabidopsis etiolated hypocotyl, the role of JAs in other model systems is still not clear.

Because JAs accumulated rapidly and transiently at the base of stem cuttings, Ahkami at al. [79] assumed that the transient JA peak at the base of *Petunia* (*Petunia hybrida*) leafy cuttings might be required for AR formation. In line with this, downregulated RNAi-*PhAOC* lines exhibited less AR compared to wild counterparts, indicating that AOC-mediated JA biosynthesis is required for AR formation in *Petunia* [77]. However, when continuously treated with OPDA, JA-Ile or MeJA, Putunia cuttings showed a decrease in AR number in a dose-dependent manner [77]. These data suggest that a certain threshold of JAs is required for ARI, but when applied continuously it inhibits this process. Therefore, further research is needed to clarify the role of JAs in ARI. The fast development of the gene-editing technologies and the availability of the first draft of *Petunia* genome will be a good opportunity to unravel the role of JAs in ARI in *Petunia*. In line with negative role of JAs in AR formation, Fattorini et al. [23] used hormone-induced organogenesis system to investigate the role of MeJA in AR formation. They reported that MeJA in combination with high dose of the natural auxin indole-3-butyric acid (IBA, 10 μM) and CK, Kinetin (Kin, 0.1 μM) promoted AR formation in Arabidopsis seedlings and stem thin cell layers when grown in continuous darkness [23, 80]. These contrasting data might be explained by the use of different experimental conditions. It is also possible that MeJA in interaction with IBA and/or Kin triggers unexpected feedback loops or signaling cascades downstream of auxin-JAs crosstalk that induced AR formation.

We reasoned that a continuous hormonal treatment with arbitrary concentrations (which might not reflect the *in planta* physiological concentrations) is not an optimum approach to investigate the role of JAs because AR formation consists of multiple and dynamic developmental steps and each step likely requires a unique signaling input. It is also likely that COI1 receptor requires precise physiological concentrations of JA-Ile to make co-receptor complexes with JAZ repressors in order to trigger specific signaling outputs. In fact, one should be careful while interpreting hormone-induced organogenesis data because using arbitrary concentrations might trigger unexpected and pleiotropic signaling circuits. In natural conditions, plants regenerate and produce new organs through very sophisticated and robust intrinsic hormonal networks.

JAs have been shown to act downstream of *ARF6* and *ARF8*, which are positive regulators and *ARF17*, which is a negative regulator of ARI. The three ARFs directly regulate the transcription of three auxin-inducible *GRETCHEN HAGEN* genes (*GH3.3*, *GH3.5*, and *GH3.6*), which control the JA pool in the hypocotyl (Fig. 1) [22, 81]. The triple mutant *gh3.3gh3.5gh3.6* accumulated more JA and JA-Ile in the hypocotyl and subsequently produced very few AR compared to wild type, suggesting that the three GH3 enzymes conjugated JA into inactive forms. This hypothesis was supported by the fact that GH3.3, GH3.5, and GH3.6 conjugated JA into amino acids (JA-Trp, JA-Asp, JA-Met) in vitro [22]. Interestingly, synthetic Jasmonoyl-L-Tryptophan (JA-Trp) is an active molecule and acts in a COI1-independent manner to interfere with root gravitropism and auxin-induced LR formation by disturbing the activity IAA influx carrier (*AUX1*) [82, 83], thus, it would be interesting to test whether endogenous JA-Trp (or other JA-conjugates) are involved in ARI. Interestingly, *ARF17* was quickly induced by exogenously applied MeJA [80], which supports the fact that JAs are a negative regulator of AR as described in Refs. [22, 81, 84].

Although the role of JAs in AR formation seems to be context- and species-dependent and further research is needed to better understand its direct target, we propose the following hypothesis for the initiation of AR at the base of stem cuttings: JA and JA-Ile rapidly and transiently accumulate at the base of cuttings forming one of the earliest wounding-associated hormonal peaks [79]. This peak is always followed by a peak of auxin [85, 86]. Auxin directly activates the expression of GH3.3, 3.5, and 3.6 enzymes, along with other cell reprogramming events [85]. The three GH3 enzymes conjugate JA to amino acids and consequently deplete the JA peak [22]. This "derepression mechanism" triggers the formative cell divisions, thereby allowing ARI.

JAs are known to strongly inhibit cell division by repressing the transcription of several cell cycle genes, including B type cyclin-dependent kinases [27, 28, 35, 87]. Similarly, wound-induced JAs have been shown to inhibit callus formation in Arabidopsis hypocotyl [88] likely by repressing cell division programs. We propose that continuous or long JA treatment inhibits AR formation likely by interfering with cell cycle progression gene expression needed for pericycle division and it might also repress the expression of genes coding for key factors required for cell fate decisions.

JA signaling pathway modulates several hormonals signaling cascades during plant growth and development and in response to environmental cues. For example, wound-induced JAs and ET antagonize each other to control many developmental processes [89]. JA-induced MYC2 is antagonized with EIN3 transcription factor and suppresses its action during apical hook formation, wound, and insect attack responses [89]. Because EIN3-mediated ET signaling has been shown to promote AR formation in Arabidopsis hypocotyl [90], it would be interesting to check MYC2-EIN3 antagonistic interaction in the context of AR formation.

In conclusion, JAs have emerged as a potential hormone controlling de novo organogenesis including adventitious rooting and further research is now needed to complete and put together different pieces of the puzzle.

10 Conclusion

To survive and adapt to the ever-changing environmental conditions, land plants have evolved sophisticated root system architecture, which can precisely perceive and quickly respond to environmental cues. Being sessile in a highly heterogeneous rhizosphere, root systems continuously grow, branch, and elongate toward water and nutrient resources and this dynamic patterning is due to its ability and plasticity to coordinate intrinsic competence and perceived cues from the surrounding microenvironment. Recent research revealed the importance of stress-associated phytohormones in mediating this process. Increasing evidences suggested that JAs might be potential signaling molecules in mediating rhizotaxis establishment. Indeed, JAs integrate several hormonal pathways to define precise cell-fate decisions for rhizotaxis establishment. Despite the great importance of the literature acquired using "artificial" growth conditions, moving toward the natural conditions, or at least to those partially mimicking natural conditions, will be the next challenge to better understand the role of different phytohormones including JAs in rhizotaxis formation. Last but not least, early land plants, such as *Marchantia polymorpha* evolved a very simple JAs biosynthesis and signaling machineries [91]; in fact, tackling the role of JAs in rhizotaxis from the *evo-devo* perspective would be exciting and will open new avenues for research on JAs.

Acknowledgments

We apologize to all colleagues whose work could not be cited due to space limitations. Research by our group is supported by the Swedish Research Council for Research and Innovation for Sustainable Growth (VINNOVA), the Swedish Research Council (VR), and the Carl Kempe Foundation.

Author Contribution: A.L. designed and wrote the manuscript with help from A.R. and C.B. A.L., C.B., and A.R. edited the manuscript.

References

1. Hesse A, Müller F (1899) Berichte d. Deutsch. Chem Gesellschaft 32:565–765
2. Demole E, Lederer E, Mercier D (1962) Isolement et détermination de la structure du jasmonate de méthyle, constituant odorant caractéristique de l'essence de jasmin. Helvetica Chimica Acta 45:675–685
3. Goossens J, Fernández-Calvo P, Schweizer F et al (2016) Jasmonates: signal transduction components and their roles in environmental stress responses. Plant Mol Biol 91:673–689
4. Howe GA, Jander G (2008) Plant immunity to insect herbivores. Annu Rev Plant Biol 59:41–66

5. Wasternack C, Hause B (2013) Jasmonates: biosynthesis, perception, signal transduction and action in plant stress response, growth and development. An update to the 2007 review in annals of botany. Ann Bot 111:1021–1058
6. Chini A, Fonseca S, Fernández G et al (2007) The JAZ family of repressors is the missing link in jasmonate signalling. Nature 448:666–671
7. Thines B, Katsir L, Melotto M et al (2007) JAZ repressor proteins are targets of the SCFCOI1 complex during jasmonate signalling. Nature 448:661–665
8. Pauwels L, Barbero GF, Geerinck J et al (2010) NINJA connects the co-repressor TOPLESS to jasmonate signalling. Nature 464:788–791
9. Shyu C, Figueroa P, DePew CL et al (2012) JAZ8 lacks a canonical degron and has an EAR motif that mediates transcriptional repression of jasmonate responses in Arabidopsis. Plant Cell Online 24:536–550
10. Sheard LB, Tan X, Mao H et al (2010) Jasmonate perception by inositol-phosphate-potentiated COI1-JAZ co-receptor. Nature 468:400–407
11. Goetz S, Hellwege A, Stenzel I et al (2012) Role of cis-12-oxo-phytodienoic acid in tomato embryo development. Plant Physiol 158:1715–1727
12. Chen J, Sonobe K, Ogawa N et al (2013) Inhibition of Arabidopsis hypocotyl elongation by jasmonates is enhanced under red light in phytochrome B dependent manner. J Plant Res 126:161–168
13. Stintzi A, Browse J (2000) The Arabidopsis male-sterile mutant, opr3, lacks the 12 oxophytodienoic acid reductase required for jasmonate synthesis. Proc Natl Acad Sci U S A 97:10625–10630
14. Park JH, Halitschke R, Kim HB et al (2002) A knock-out mutation in allene oxide synthase results in male sterility and defective wound signal transduction in Arabidopsis due to a block in jasmonic acid biosynthesis. Plant J 31:1–12
15. Caldelari D, Wang G, Farmer EE et al (2011) Arabidopsis lox3 lox4 double mutants are male sterile and defective in global proliferative arrest. Plant Mol Biol 75:25–33
16. Zhai Q, Zhang X, Wu F et al (2015) Transcriptional mechanism of jasmonate receptor COI1-mediated delay of flowering time in Arabidopsis. Plant Cell 27:2814–2828
17. Shan X, Wang J, Chua L et al (2011) The role of Arabidopsis rubisco activase in jasmonate-induced leaf senescence. Plant Physiol 155:751–764
18. Noir S, Bomer M, Takahashi N et al (2013) Jasmonate controls leaf growth by repressing cell proliferation and the onset of endoreduplication while maintaining a potential stand-by mode. Plant Physiol 161:1930–1951
19. Feys B, Benedetti CE, Penfold CN et al (1994) Arabidopsis mutants selected for resistance to the phytotoxin coronatine are male sterile, insensitive to methyl jasmonate, and resistant to a bacterial pathogen. Plant Cell Online 6:751–759
20. Staswick PE, Sut W, Howell SH (1992) Methyl jasmonate inhibition of root growth and induction of a leaf protein are decreased in an Arabidopsis thaliana mutant. Proc Natl Acad Sci U S A 89:6837–6840
21. Vellosillo T, Martinez M, Lopez MA et al (2007) Oxylipins produced by the 9-lipoxygenase pathway in Arabidopsis regulate lateral root development and defense responses through a specific signaling cascade. Plant Cell Online 19:831–846
22. Gutierrez L, Mongelard G, Floková K et al (2012) Auxin controls *Arabidopsis* adventitious root initiation by regulating jasmonic acid homeostasis. Plant Cell 24:2515–2527
23. Fattorini L, Falasca G, Kevers C et al (2009) Adventitious rooting is enhanced by methyl jasmonate in tobacco thin cell layers. Planta 231:155–168
24. Petricka JJ, Winter CM, Benfey PN (2012) Control of *Arabidopsis* root development. Ann Rev Plant Biol 63:563–590
25. Dathe W, Rönsch H, Preiss A et al (1981) Endogenous plant hormones of the broad bean, Vicia faba L. (-)-jasmonic acid a plant growth inhibitor in pericarp. Planta. 155:530–535
26. Swiatek A, Van Dongen W, Esmans EL et al (2004) Metabolic fate of jasmonates in tobacco bright yellow-2 cells. Plant Physiol 135:161–172
27. Pauwels L, Morreel K, De Witte E et al (2008) Mapping methyl jasmonate-mediated transcriptional reprogramming of metabolism and cell cycle progression in cultured Arabidopsis cells. Proc Natl Acad Sci U S A 105:1380–1385
28. Chen Q, Sun J, Zhai Q et al (2011) The basic helix-loop-helix transcription factor MYC2 directly represses PLETHORA expression during jasmonate-mediated modulation of the root stem cell niche in Arabidopsis. Plant Cell Online 23:3335–3352
29. Aida M, Beis D, Heidstra R et al (2004) The PLETHORA genes mediate patterning of the

Arabidopsis root stem cell niche. Cell 119:119–120
30. Chini A, Gimenez-Ibanez S, Goossens A et al (2016) Redundancy and specificity in jasmonate signalling. Curr Opin Plant Biol 33:147–156
31. Thireault C, Shyu C, Yoshida Y et al (2015) Repression of jasmonate signaling by a non-TIFY JAZ protein in Arabidopsis. Plant J 82:669–679
32. Campos ML, Yoshida Y, Major IT et al (2016) Rewiring of jasmonate and phytochrome B signalling uncouples plant growth-defense tradeoffs. Nat Commun 7:1–10
33. Guo Q, Yoshida Y, Major IT et al (2018) JAZ repressors of metabolic defense promote growth and reproductive fitness in *Arabidopsis*. Proc Natl Acad Sci U S A 115:10768–10777
34. Acosta IF, Gasperini D, Chételat A, Stolz S, Santuari L, EEF (2013) Role of NINJA in root jasmonate signaling. Proc Natl Acad Sci U S A 110:15473–15478
35. Gasperini D, Chételat A, Acosta IF et al (2015) Multilayered organization of jasmonate signalling in the regulation of root growth. PLoS Genet 11:1–27
36. Fernández-Calvo P, Chini A, Fernández-Barbero G et al (2011) The *Arabidopsis* bHLH transcription factors MYC3 and MYC4 are targets of JAZ repressors and act additively with MYC2 in the activation of jasmonate responses. Plant Cell 23:701–715
37. Bömer M, O'Brien JA, Pérez-Salamó I et al (2018) COI1-dependent jasmonate signalling affects growth, metabolite production and cell wall protein composition in Arabidopsis. Ann Bot 31:1117–1129
38. Gasperini D, Chauvin A, Acosta IF et al (2015) Axial and radial oxylipin transport. Plant Physiol 169:2244–2254
39. Li Q, Zheng J, Li S et al (2017) Transporter-mediated nuclear entry of jasmonoyl-isoleucine is essential for jasmonate signaling. Mol Plant 10:695–708
40. Sun J, Chen Q, Qi L et al (2011) Jasmonate modulates endocytosis and plasma membrane accumulation of the Arabidopsis pin2 protein. New Phytol 191:360–375
41. Leyser HMO, Lincoln CA, Timpte C et al (1993) Arabidopsis auxin-resistance gene AXR1 encodes a protein related to ubiquitin-activating enzyme E1. Nature 364:161–164
42. Dharmasiri N, Dharmasiri S, Estelle M (2005) The F-box protein TIR1 is an auxin receptor. Nature 435:441–445
43. Kepinski S, Leyser O (2005) The Arabidopsis F-box protein TIR1 is an auxin receptor. Nature 435:446–451
44. Stepanova AN, Hoyt JM, Hamilton AA et al (2005) A link between ethylene and auxin uncovered by the characterization of two root-specific ethylene-insensitive mutants in Arabidopsis. Plant Cell Online 17:2230–2242
45. Zhu Z, An F, Feng Y et al (2011) Derepression of ethylene-stabilized transcription factors (EIN3/EIL1) mediates jasmonate and ethylene signaling synergy in Arabidopsis. Proc Natl Acad Sci U S A 108:12539–12544
46. Mao JL, Miao ZQ, Wang Z et al (2016) Arabidopsis ERF1 mediates cross-talk between ethylene and auxin biosynthesis during primary root elongation by regulating ASA1 expression. PLoS Genet 12:1–20
47. Cai XT, Xu P, Zhao PX et al (2014) Arabidopsis ERF109 mediates cross-talk between jasmonic acid and auxin biosynthesis during lateral root formation. Nat Commun 5:1–13
48. Wei Z, Li J (2016) Brassinosteroids regulate root growth, development, and symbiosis. Mol Plant 9:86–100
49. Ren C, Han C, Peng W et al (2009) A leaky mutation in DWARF4 reveals an antagonistic role of brassinosteroid in the inhibition of root growth by jasmonate in Arabidopsis. Plant Physiol 151:1412–1420
50. Kim H, Kwon M, Ryu H et al (2006) The regulation of DWARF4 expression is likely a critical mechanism in maintaining the homeostasis of bioactive brassinosteroids in Arabidopsis. Plant Physiol 140:548–557
51. Kim B, Fujioka S, Kwon M et al (2013) Arabidopsis Brassinosteroid-overproducing gulliver3-D/dwarf4-D mutants exhibit altered responses to Jasmonic acid and pathogen. Plant Cell Rep 32:1139–1149
52. Avalbaev A, Yuldashev R, Fedorova K et al (2016) Exogenous methyl jasmonate regulates cytokinin content by modulating cytokinin oxidase activity in wheat seedlings under salinity. J Plant Physiol 191:101–110
53. Jang G, Chang SH, Um TY et al (2017) Antagonistic interaction between jasmonic acid and cytokinin in xylem development. Sci Rep 7:1–13
54. Yang Z-B, He C, Ma Y et al (2017) Jasmonic acid enhances Al-induced root growth inhibition. Plant Physiol 173:1420–1433
55. Van Norman JM, Xuan W, Beeckman T et al (2013) To branch or not to branch: the role of pre-patterning in lateral root formation. Development 140:4301–4310
56. Ötvös K, Benková E (2017) Spatiotemporal mechanisms of root branching. Curr Opin Genet Dev 45:82–89

57. Bellini C, Pacurar DI, Perrone I (2014) Adventitious roots and lateral roots: similarities and differences. Ann Rev Plant Biol 65:639–666
58. Grunewald W, Vanholme B, Pauwels L et al (2009) Expression of the Arabidopsis jasmonate signalling repressor JAZ1/TIFY10A is stimulated by auxin. EMBO Rep 10:923–928
59. Sun J, Xu Y, Ye S et al (2009) Arabidopsis ASA1 is important for jasmonate-mediated regulation of auxin biosynthesis and transport during lateral root formation. Plant Cell Online 21:1495–1511
60. Wang SC, Ichii M, Taketa S et al (2002) Effect of jasmonic acid on lateral root formation in rice seedling. Acta Bot Sin 44:502–504
61. Hentrich M, Böttcher C, Düchting P et al (2013) The jasmonic acid signaling pathway is linked to auxin homeostasis through the modulation of YUCCA8 and YUCCA9 gene expression. Plant J 74:626–637
62. Raya-González J, Pelagio-Flores R, López-Bucio J (2012) The jasmonate receptor COI1 plays a role in jasmonate-induced lateral root formation and lateral root positioning in Arabidopsis thaliana. J Plant Physiol 169:1348–1358
63. Yadav V, Mallappa C, Gangappa SN et al (2005) A basic helix-loop-helix transcription factor in Arabidopsis, MYC2, acts as a repressor of blue light–mediated photomorphogenic growth. Plant Cell 17:1953–1966
64. Gangappa SN, Prasad VBR, Chattopadhyay S (2010) Functional interconnection of MYC2 and SPA1 in the photomorphogenic seedling development of Arabidopsis. Plant Physiol 154:1210–1219
65. Gangappa SN, Chattopadhyay S (2010) MYC2, a bHLH transcription factor, modulates the adult phenotype of SPA1. Plant Signal Behav 5:1650–1652
66. Ishimaru Y, Hayashi K, Suzuki T et al (2018) Jasmonic acid inhibits auxin-induced lateral rooting independently of the CORONATINE INSENSITIVE 1 receptor. Plant Physiol 177:1704–1716
67. Morris EC, Griffiths M, Golebiowska A et al (2017) Shaping 3D root system architecture. Curr Biol 27:919–930
68. Ditengou FA, Teale WD, Kochersperger P et al (2008) Mechanical induction of lateral root initiation in Arabidopsis thaliana. Proc Natl Acad Sci U S A 105:18818–18823
69. Sheng L, Hu X, Du Y, et al (2017) Non-canonical WOX11-mediated root branching contributes to plasticity in Arabidopsis root system architecture. Development. dev.152132
70. Chehab EW, Yao C, Henderson Z et al (2012) Arabidopsis touch-induced morphogenesis is jasmonate mediated and protects against pests. Curr Biol 22:701–706
71. Guo H, Nolan TM, Song G et al (2018) FERONIA receptor kinase contributes to plant immunity by suppressing jasmonic acid signaling in Arabidopsis thaliana. Curr Biol 28:3316–3324.e6
72. Dong Q, Zhang Z, Liu Y et al (2019) FERONIA regulates auxin-mediated lateral root development and primary root gravitropism. FEBS Lett 593:97–106
73. Gonneau M, Desprez T, Martin M et al (2018) Receptor kinase THESEUS1 is a rapid alkalinization factor 34 receptor in Arabidopsis. Curr Biol 28:2452–2458.e4
74. Tardieu F, Cabrera-Bosquet L, Pridmore T et al (2017) Plant phenomics, from sensors to knowledge. Curr Biol 27:R770–R783
75. Larrieu A, Champion A, Legrand J et al (2015) A fluorescent hormone biosensor reveals the dynamics of jasmonate signalling in plants. Nat Commun 6:1–9
76. Lakehal A, Bellini C (2018) Control of adventitious root formation: insights into synergistic and antagonistic hormonal interactions. Physiol Plant 165:90–100
77. Lischweski S, Muchow A, Guthörl D et al (2015) Jasmonates act positively in adventitious root formation in petunia cuttings. BMC Plant Biol 15:1–10
78. Chen LR, Chen YJ, Lee CY et al (2007) MeJA-induced transcriptional changes in adventitious roots of Bupleurum kaoi. Plant Sci 173:12–24
79. Ahkami AH, Lischewski S, Haensch KT et al (2009) Molecular physiology of adventitious root formation in Petunia hybrida cuttings: involvement of wound response and primary metabolism. New Phytol 181:613–625
80. Fattorini L, Hause B, Gutierrez L et al (2018) Jasmonate promotes auxin-induced adventitious rooting in dark-grown Arabidopsis thaliana seedlings and stem thin cell layers by a cross-talk with ethylene signalling and a modulation of xylogenesis. BMC Plant Biol 18:1–18
81. Gutierrez L, Bussell JD, Pacurar DI et al (2009) Phenotypic plasticity of adventitious rooting in Arabidopsis is controlled by complex regulation of AUXIN RESPONSE FACTOR transcripts and MicroRNA abundance. Plant Cell 21:3119–3132
82. Staswick P, Rowe M, Spalding EP et al (2017) Jasmonoyl-L-Tryptophan Disrupts IAA activity through the AUX1 Auxin Permease. Front Plant Sci 8:1–10

83. Staswick PE (2009) The tryptophan conjugates of jasmonic and indole-3-acetic acids are endogenous auxin inhibitors. Plant Physiol 150:1310–1321
84. Sorin C, Bussell JD, Camus I et al (2005) Auxin and light control of adventitious rooting in *Arabidopsis*. Plant Cell 17:1343–1359
85. Ahkami AH, Melzer M, Ghaffari MR et al (2013) Distribution of indole-3-acetic acid in Petunia hybrida shoot tip cuttings and relationship between auxin transport, carbohydrate metabolism and adventitious root formation. Planta 238:499–517
86. Rasmussen A, Hosseini SA, Hajirezaei MR et al (2015) Adventitious rooting declines with the vegetative to reproductive switch and involves a changed auxin homeostasis. J Exp Bot 66:1437–1452
87. Swiatek A, Lenjou M, Van Bockstaele D et al (2002) Differential effect of jasmonic acid and abscisic acid on cell cycle progression in tobacco BY-2 cells. Plant Physiol 128:201–211
88. Ikeuchi M, Iwase A, Rymen B et al (2017) Wounding triggers callus formation via dynamic hormonal and transcriptional changes. Plant Physiol 175:1158–1174
89. Song S, Huang H, Gao H et al (2014) Interaction between MYC2 and ETHYLENE INSENSITIVE3 modulates antagonism between jasmonate and ethylene signaling in Arabidopsis. Plant Cell 26:263–279
90. Rasmussen A, Hu Y, Depaepe T et al (2017) Ethylene controls adventitious root initiation sites in Arabidopsis hypocotyls independently of strigolactones. J Plant Growth Regul 36:897–911
91. Bowman JL, Kohchi T, Yamato KT et al (2017) Insights into land plant evolution garnered from the marchantia polymorpha genome. Cell 171:287–304.e15
92. Sukumar P, Maloney GS, Muday GK (2013) Localized induction of the ATP-binding cassette B19 auxin transporter enhances adventitious root formation in Arabidopsis. Plant Physiol 162:1392–1405
93. Blilou I, Xu J, Wildwater M et al (2005) The PIN auxin efflux facilitator network controls growth and patterning in Arabidopsis roots. Nature 433:39–44

Chapter 2

Phenotyping of Light Response on JA-Defective Mutant in Rice

Michael Riemann

Abstract

Jasmonates are associated with stress responses. However, there are many roles for jasmonates in plant development. One such function is their participation in the regulation of growth and development of rice seedlings, especially of their coleoptile in response to light. Impairment of jasmonate biosynthesis or signaling results in obvious phenotypic effects in seedlings, which can be used for screening homozygous mutants from a heterozygous population for downstream experimental approaches. A procedure for screening mutants under light is described, which can easily be adapted to standard laboratory conditions.

Key words Jasmonate rice, Coleoptile, *Oryza sativa*, Photomorphogenesis, Monochromatic light, Mutant

1 Introduction

Apart from being an important mediator in stress responses, jasmonates are regulators of physiological and developmental processes [1]. For example, jasmonates control sex determination in maize [2], tendril coiling [3], or senescence [4]. One developmental function of jasmonate in rice is the control of seedling growth. Especially, elongation growth of the coleoptile, a specialized tissue protecting the leaves of rice seedlings from mechanical constraints during growth in the soil, is a jasmonate-dependent process. This was found through mutant approaches in which mutants were identified showing elongated coleoptiles under monochromatic red light [5, 6]. These mutants have been shown to be impaired in the functions of ALLENE OXIDE 1 (AOS1, [7]) and ALLENE OXIDE CYCLASE (AOC, [8]), two key enzymes of jasmonate biosynthesis. Also, mutations in the rice ortholog of JASMONATE RESISTANT1, an enzyme conjugation the amino acid isoleucine to jasmonic acid [9], cause an elongated coleoptile phenotype in light, although the phenotype appears more subtle as compared to *aos1* or *aoc* mutants [10, 11]. How jasmonates influence the

Antony Champion and Laurent Laplaze (eds.), *Jasmonate in Plant Biology: Methods and Protocols*, Methods in Molecular Biology, vol. 2085, https://doi.org/10.1007/978-1-0716-0142-6_2,

elongation growth is not fully clarified, [12] but it has been found that they also feedback to the photoreceptors of the phytochrome family [13], for example, by influencing their stability [14]. Cryptochromes, blue light photoreceptors, also likely influence JA biosynthesis as overexpression of *OsCRY1* led to enhanced levels of *OsAOS1* transcripts [15]. Unfortunately, impairing jasmonate biosynthesis in rice also leads to strongly reduced fertility [8, 10, 16], which makes it impossible to maintain homozygous lines in some cases or reduces the amount of homozygous mutant seeds significantly. Therefore, it becomes inevitable to work with heterozygous mutant populations. In order to select homozygous mutant plants from such a heterozygous population, DNA extraction and PCR analysis are required. These are time- and labor-intensive steps and sometimes downstream experiments might require minimal manipulation of the plants. As the seedling phenotypes of homozygous mutant plants are very clear and appear early in development (Fig. 1), selection based on phenotypic traits of seedlings is a powerful way to conduct experiments with homozygous jasmonate mutant plants. Basically, selection works in plants grown both in darkness and in light. As photobiological darkness is not easy to obtain, the procedure is mainly described for light-grown plants here.

Fig. 1 Phenotypic differentiation of wild type (cv. Nihonmasari) and jasmonate biosynthesis mutants (*coleoptile photomorphogenesis 2, cpm2*). Seedlings were raised for 7 days at 25 °C in darkness (**a**) or continuous blue light (**b**, 10 μmol m^{-2} s^{-1}). Arrows indicate the position of the node between coleoptile and mesocotyl. (**a**) Etiolated wild-type seedlings develop a long coleoptile. Seedlings of the *AOC* mutant *cpm2* can easily be distinguished from the wild type by an elongated mesocotyl. (**b**) Coleoptiles of wild-type seedlings raised in continuous light remain short, while mutant coleoptiles are significantly longer. Size bar = 10 mm

2 Materials

2.1 Plant Cultivation

1. Sterilize growth vessels (Magenta vessel GA-7 or similar). The vessels should be transparent in order to allow screening in light.
2. Prepare 0.4% phytoagar (Duchefa) and pour 100 ml (volume needed for 1 Magenta box) into sterilized growth vessels.
3. Autoclave water (240 ml per seed batch required).
4. Seed sterilization is done in 50 ml screw cap tubes. Each tube can accommodate up to 150 rice seeds. One such tube is defined as seed batch in this description. Prepare 70% of ethanol (abs.; 10 ml per seed batch), and a sodium hypochlorite solution containing around 5% of active chlorine freshly (40 ml per seed batch).
5. To raise the seedlings a standard plant growth cabinet is required. Raise the plants either in continuous light (120 μmol $m^{-2}s^{-1}$) or in a day–night cycle. Temperature should be adjusted between 25 and 28 °C and not below 20 °C if night cycles are used.
6. Optional: In order to obtain proper dark conditions, seedlings should be raised in a light tight box and covered with light tight material (e.g., a black cloth). This box should be placed in a cupboard in a dark room at 25–28 °C, ideally with no light source and window. Seedlings are extremely sensitive to light and will detect minimal amounts of light.

2.2 DNA Extraction

1. 1.5% CTAB: 9.08 g/l Tris, 15 g/l cetyltrimethylammonium bromide, 5.58 g/l disodium EDTA dihydrate (pH8.0), 61.36 g/l NaCl (pH 8.0).
2. (24:1 v/v) Chloroform:isoamyl alcohol.
3. TE buffer (pH 8.0): 12.11 g/l Tris, 3.72 g/l disodium EDTA dihydrate (pH 8.0); preparation according to [17].
4. Isopropanol (cooled at −20 °C).
5. 70% ethanol.
6. RNAse A.

3 Methods

3.1 Seed Sterilization and Plant Growth

1. Dehusk rice seeds carefully. Transfer up to 150 seeds into a 50-ml screw-cap tube.
2. Submerge seeds with 10 ml of 70 ml ethanol (abs.) and move tube horizontally for 1 min. Make sure that ethanol flows properly around the seeds.

3. Remove ethanol and wash 2 times with 40 ml water (*see* **Note 1**).
4. Remove water and submerge seeds into 40 ml of sodium hypochlorite solution (5% active chlorine). Shake horizontally for 20 min at approximately 100 rpm. Adjust the speed of the shaker such that seeds are moving slightly and sodium hypochlorite flows around the seeds properly.
5. Remove sodium hypochlorite solution under a fume hood in special waste containers. Next steps will take place under a clean bench.
6. Wash seeds four times with water.
7. Remove water. Up to 30 seeds could be sown in each Magenta box. It is important to place them equidistantly into the agar. The seeds should not be submerged into the agar, but remain at the surface. The embryo has to face upwards.
8. Incubate the plants under either light or dark condition. Phenotypes will be obvious after 5–7 days depending on the temperature conditions.
9. After this period, seedlings can be carefully removed from the agar and selected based on the phenotype. In light conditions, homozygous jasmonate biosynthesis mutants have significantly longer coleoptiles. Seedlings grown in darkness have a very prominent, elongated mesocotyl (*see* **Note 2**).
10. For quantitative analysis: In order to record a growth distribution of a heterozygous population, seedlings should be scanned with a standard flatbed scanner. Place a scale on each scan.
11. Using ImageJ (https://imagej.nih.gov/ij/), the length of coleoptiles can be measured. Count the number of coleoptiles in each 1 mm interval (1–2 mm, 2–3 mm, 3–4 mm, etc.) and depict these intervals versus the number of coleoptiles. Homozygous jasmonate biosynthesis mutants should be grouped into a subpopulation with significantly longer coleoptiles compared to homozygous wild-type and heterozygous plants, if it is a recessive gene [8].

3.2 DNA Extraction

To confirm the robustness of the phenotypic selection, parts of the plants that are not required for the downstream analysis can be harvested at the end of the experiments and used for DNA extraction. An efficient protocol for DNA extraction is suggested below.

1. Cut part of a leaf (approx. 2 cm) into small pieces, freeze in liquid nitrogen and grind leaf tissue to powder. Transfer the powder into a 2 ml reaction tube.
2. Add 900 μl boiled 1.5% CTAB buffer and incubate for 1 h at 65 °C.

3. Add 630 μl of chloroform:isoamyl alcohol solution, and shake horizontally for 15 min at 75 rpm at room temperature.
4. Centrifuge at 17,000 × *g* in a table centrifuge for 10 min.
5. Transfer the supernatant containing the nucleic acids into a new 2 ml reaction tube and precipitate DNA with 2/3 v/v ice-cold isopropanol (e.g., supernatant 700 μl, isopropanol 450 μl).
6. Shake gently and centrifuge at 17,000 × *g* for additional 10 min. Remove the supernatant (be careful to keep the pellet).
7. Wash the precipitated nucleic acids with 1 ml 70% ethanol, centrifuge and remove supernatant (be careful to keep the pellet).
8. Dry in a vacuum centrifuge for 15 min.
9. Dissolve DNA in 50 μl 1/10 TE buffer containing 5 μg RNase A for RNA digestion. Keep the sample for 2 h at room temperature before deep-freezing at −20°C. This DNA can be used for subsequent genotyping by PCR with suitable primers.

4 Notes

1. In order to discard liquids during the seed sterilization procedure, the lid can be put on the opening of the tube obliquely to prevent seeds from being flushed out (*see* Subheading 3.1, **steps 3**–7).
2. Basically, phenotypic selection can be used as a reliable method to choose jasmonate-deficient plants (*see* Subheading 3.1, **steps 9–11**). However, in cases with less clear phenotypes (e.g., *jar1* mutants), PCR-based genotyping might be necessary. In case of *jar1* rice mutants, screening in monochromatic blue or far-red light will result in a clearer phenotypic differentiation.

References

1. Wasternack C, Hause B (2013) Jasmonates: biosynthesis, perception, signal transduction and action in plant stress response, growth and development. An update to the 2007 review in Annals of Botany. Ann Bot 111:1021–1058. https://doi.org/10.1093/aob/mct067
2. Acosta IF, Laparra H, Romero SP et al (2009) Tasselseed1 is a lipoxygenase affecting jasmonic acid signaling in sex determination of maize. Science 323:262–265. https://doi.org/10.1126/science.1164645
3. Falkenstein E, Groth B, Mithöfer A, Weiler E (1991) Methyljasmonate and α-linolenic acid are potent inducers of tendril coiling. Planta 185:316–322. https://doi.org/10.1007/BF00201050
4. Yan Y, Christensen S, Isakeit T et al (2012) Disruption of *OPR7* and *OPR8* reveals the versatile functions of jasmonic acid in maize development and defense. Plant Cell. https://doi.org/10.1057/9780230524545
5. Biswas KK, Neumann R, Haga K et al (2003) Photomorphogenesis of rice seedlings: a mutant impaired in phytochrome-mediated inhibition of coleoptile growth. Plant Cell Physiol 44:242–254

6. Riemann M, Muller A, Korte A et al (2003) Impaired induction of the jasmonate pathway in the rice mutant hebiba. Plant Physiol 133:1820–1830. https://doi.org/10.1104/pp.103.027490
7. Haga K, Iino M (2004) Phytochrome-mediated transcriptional up-regulation of ALLENE OXIDE SYNTHASE in rice seedlings. Plant Cell Physiol 45:119–128. https://doi.org/10.1093/pcp/pch025
8. Riemann M, Haga K, Shimizu T et al (2013) Identification of rice allene oxide cyclase mutants and the function of jasmonate for defence against Magnaporthe oryzae. Plant J 74:226–238. https://doi.org/10.1111/tpj.12115
9. Staswick PE, Tiryaki I, Rowe ML (2002) Jasmonate response locus JAR1 and several related Arabidopsis genes encode enzymes of the firefly luciferase superfamily that show activity on jasmonic, salicylic, and indole-3-acetic acids in an assay for adenylation. Plant Cell 14:1405–1415
10. Riemann M, Riemann M, Takano M (2008) Rice JASMONATE RESISTANT 1 is involved in phytochrome and jasmonate signalling. Plant Cell Environ 31:783–792. https://doi.org/10.1111/j.1365-3040.2008.01790.x
11. Svyatyna K, Jikumaru Y, Brendel R et al (2014) Light induces jasmonate-isoleucine conjugation via OsJAR1-dependent and -independent pathways in rice. Plant Cell Environ 37:827–839. https://doi.org/10.1111/pce.12201
12. Svyatyna K, Riemann M (2012) Light-dependent regulation of the jasmonate pathway. Protoplasma 249:137–145. https://doi.org/10.1007/s00709-012-0409-3
13. Brendel R, Svyatyna K, Jikumaru Y et al (2014) Effects of light and wounding on jasmonates in rice phyAphyC mutants. Plants 3:143–159. https://doi.org/10.3390/plants3010143
14. Riemann M, Bouyer D, Hisada A et al (2009) Phytochrome A requires jasmonate for photodestruction. Planta 229:1035–1045. https://doi.org/10.1007/s00425-009-0891-9
15. Hirose F, Shinomura T, Tanabata T et al (2006) Involvement of rice cryptochromes in de-etiolation responses and flowering. Plant Cell Physiol 47(7):915–925. https://doi.org/10.1093/pcp/pcj064
16. Dhakarey R, Peethambaran PK, Riemann M (2016) Functional analysis of jasmonates in rice through mutant approaches. Plants 5(15): E15
17. Sambrock J, Fritsch EF, Maniatis T (1989) Molecular cloning, a laboratory manual, 2nd edn, vols 1, 2 and 3. Cold Spring Harbor Laboratory Press, Cold Spring Harbor, NY

Chapter 3

Techniques to Determine the Effects of Jasmonates on Root Hydraulic Conductivity

Beatriz Sánchez-Romera and Ricardo Aroca

Abstract

Plants subjected to drought and saline stress conditions suffer from tissue dehydration. Such dehydration is caused by the imbalance between root water uptake by roots and water loss by transpiration. Therefore, determination of root hydraulic properties is crucial to understand plant water balance. Root hydraulic conductivity (L) can be used to estimate root water transport capacity. L depends on root architecture (length and diameter of the root and proliferation of secondary roots), radial water transport pathway (root xylem vessels, plasmodesmata, apoplastic space, caspary bands), and on intrinsic membrane permeability to water (aquaporins, water membrane protein channels). Different methods have been developed to measure L, such as Pressure Chamber, Free Exudation, High-Pressure Flowmeter (HPFM), and Root Pressure Probe (RPP). In this chapter, we will focus on Pressure Chamber, Free Exudation, and HPFM methods which have been used to determine the effect of jasmonates (JA) on root hydraulic conductivity.

Key words Exudate, Flow, Hydraulic conductivity, Osmolality, Pressure, Root, Water, Xylem

1 Introduction

Root hydraulic conductivity (L) is an intrinsic characteristic that determines root water transport capacity. Determination of L is very useful because it can be used to know the contribution of the different water pathways across roots (apoplastic versus symplastic) and indicates plant water use strategies to cope with different stress situations where water is a limited resource, such as drought and salinity [1, 2].

In this chapter, we describe the free exudate, pressure chamber and High-Pressure Flowmeter (HPFM) methods because they have been used in our research group to quantify the role of JA in regulating L. These methods are used in several research groups specialized in the study of plant water relations and are supported by numerous scientific publications. Although we explain these methods step-by-step, slight modifications might be found to adapt the method to different plant species [3, 4].

Antony Champion and Laurent Laplaze (eds.), *Jasmonate in Plant Biology: Methods and Protocols*, Methods in Molecular Biology, vol. 2085, https://doi.org/10.1007/978-1-0716-0142-6_3,

Fig. 1 (**a**) Pressure chamber (**b**) free exudation method in plants grown under hydroponic conditions

1. Pressure Chamber Method: It is the first method that was described to quantify L [5] and the most popular. It consists of a metal chamber connected to a compressed air tank where the root system is introduced into known solution (Fig. 1a). The stem base has to be perfectly sealed with plastic disk adapters at the top of the chamber. Then, the root is subjected to different pressures for a limited amount of time and the exudate, which flows from the xylem vessels during that time, is collected. In other groups, the stem is directly connected to a water-flow sensor making measures faster and easier.

   L measured by pressure chamber method is indicative of water movements through apoplastic routes (water movement through intercellular spaces) and cell-to-cell routes, that is, plasmodesmata and protein channels located in cell membranes, known as aquaporins.

   (a) Advantages: this method is very effective and fast. It is recommended for plants with low L.

   (b) Disadvantages: depending on the size of the plant, sometimes the stem base does not fit into the metal chamber. It is a technique that requires working with a compressed air tank that could be dangerous.

2. Free Exudation Method: Osmotic root hydraulic conductivity is estimated using this method (Fig. 1b). According to Steudle's model [6], water flows mostly through the cell-to-cell pathway, because detached root system exudates at atmospheric pressure due to stomatal conductance have been eliminated, which is related to movement of water by an apoplastic pathway. Under these conditions, water circulates through roots following an osmotic gradient between the root bathing solution and the root xylem vessels. We recommend carrying out an initial screening using this method before starting a new research project, since it provides reliable L information and can be used to analyze many plants simultaneously.

(a) Advantages: it is easy to use, does not require great technical skill. It does not involve any dangerous material.

(b) Disadvantages: some chemical treatments, salinity and severe drought induce very low levels of L that do not allow the collection of exudates. Therefore, it is recommended for use in plants with high hydraulic conductivity since the amount of the exudate that is collected is very small. In addition, it is necessary in an osmometer to measure osmolarity of the exudate and water solution.

3. High Pressure Flow Meter Method (HPFM): HPFM was described by Tyree et al. [7, 8]. This method is characterized by the application of different flows of water from the stem to the root of the plant during a limited time (Fig. 2). In this method, the flow of water is in the opposite direction compared to Free Exudate and Pressure Chamber method. This method quantifies the movement of water through the apoplastic route and cell-to-cell. It is recommended to use on large plants like trees, where it is not necessary to remove the plant from the pot and thus not to manipulate the root.

   (a) Advantages: it is not necessary to manipulate the plant root system and reduce the effects of hydraulic capacitance in roots and shoots of small plants.

   (b) Disadvantages: the water intake system to the plant is very sensitive, small air bubbles can cause errors.

2 Materials

2.1 Pressure Chamber Method

1. A pressure chamber formed by one pressure regulator and three chambers (in our case) with double lid (Fig. 1a).
2. A razor blade to cut the aerial part.
3. Water or nutritional solution has to be included inside the chambers.
4. Adapter rings to prevent air loss around the stem. There are some adapters that must be placed on the stem so that the pressure chamber is completely hermetic (Fig. 3a). There are several models with different diameters. The adapter, that best fits the stem of the plant, will be selected.
5. A silicon tube to collect the exudates. We design these in different sizes to better adjust to the stem. The plastic tube is formed by a piece of silicone tube and a pipette tip. The top and the base of pipette tip are cut with scissors to enable a good adaptability. We take advantage of the conical shape of the pipette tip to adapt it to different stem diameters (Fig. 3b).

Fig. 2 High pressure flow meter

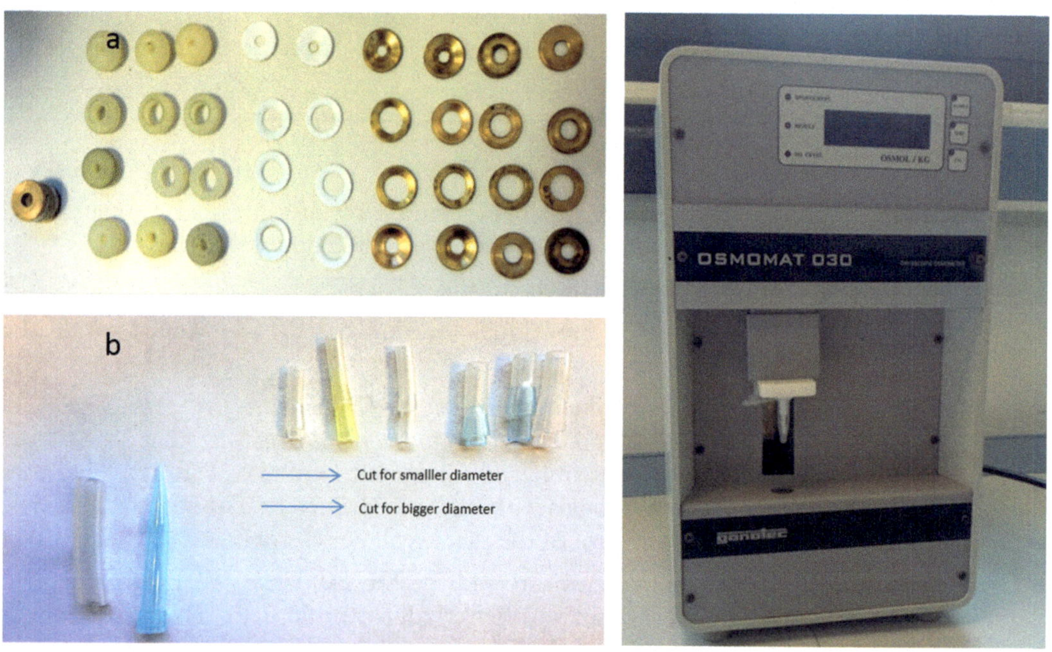

Fig. 3 (**a**) Adapter rings to fix the plant to the pressure chamber, (**b**) Plastic tube to collect the exudates, and (**c**) osmometer

6. Eppendorf tubes to collect the total exudate of the plant, previously weighed.
7. A balance to weigh the root dry weight and the exudates collected.
8. A timer to measure the time of exudation.
9. An oven to dry the root samples.
10. A compressed air tank.

2.2 Free Exudation Method

1. A razor blade to cut the aerial part.
2. Silicon tubes to collect exudates (Fig. 3b).
3. Eppendorf tubes to collect total exudate of the plant, previously weighed.
4. Trays or containers to saturate soil in case plants are grown in pots.
5. Air pumps to enhance water aeration, since the plants are immersed in solution and L results could be affected by hypoxic conditions.
6. An osmometer, this machine is used to quantify the concentration of osmolites in the exudates and water of the trays (Fig. 3c).
7. A balance to weigh root dry weight and exudates collected.
8. A timer to measure the time of exudation.
9. An oven to dry the root samples.

2.3 HPFM

1. A HPFM machine (Fig. 3).
2. A razor blade to cut the aerial part.
3. A plastic chamber where the plant stem is connected (Fig. 4a).
4. Pressure couplings (Fig. 4b).
5. A balance to weigh root dry weight.
6. An oven to dry root samples.
7. A computer with software (HPFM-3, which is included when the machine is bought).
8. A compressed air tank.

3 Methods

3.1 Pressure Chamber Method

1. L of roots is measured by pressurizing the roots in a pressure chamber.
2. The stems of plants are excised with a blade, 3 cm above the root (*see* **Note 1**) (Fig. 5a).

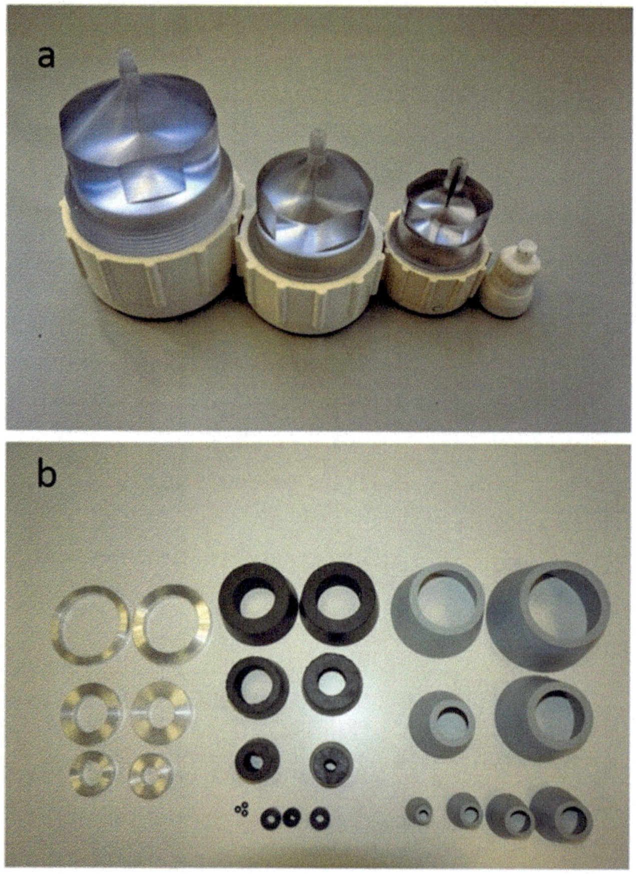

Fig. 4 Pressure couplings for HPFM. (**a**) Chamber where the plant and adapters are inserted and (**b**) adapter rings which are inserted inside the chamber

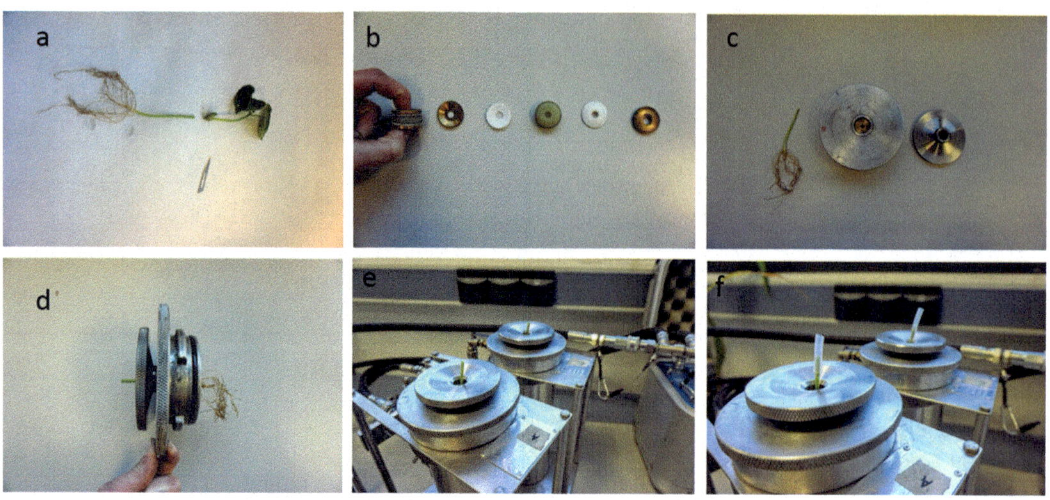

Fig. 5 Sample preparation: (**a**) cutting of the stem, (**b**) adapter ring position, (**c**) and (**d**) assembly of the plant in the lid with adapter rings, (**e**) plant inside the chamber and (**f**) stem with a collector tube

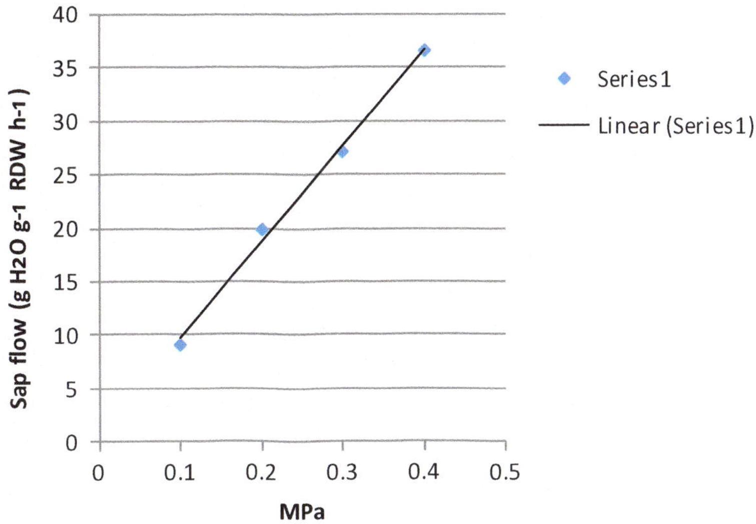

Fig. 6 Example of L determination by the pressure chamber method (real example). R2 close to 1 depicts a good use of the method. L value is 89,711 g H_20 g^{-1} Root Dry Weight (RDW) MPa^{-1} h^{-1}

3. Select the adaptors that have a diameter to similar the stem (Fig. 5b).
4. Put the adaptors following the same order than Fig. 5b. inside two piece of the lip (Fig. 5c).
5. Introduce the stem into the hole of the lid (Fig. 5d).
6. Root systems and lid are placed inside the pressure chamber containing a nutrient solution and chemicals for different treatments (Fig. 5e).
7. Silicon tubes (Fig. 5b) are connected to the stems to collect exudates as shown in Fig. 6f (*see* **Note 2**).
8. Finally, roots are subjected to different pressures (0.2, 0.3, 0.4, and 0.5 MPa) for 3 min at each pressure (*see* **Note 3**).
9. The exuded sap is collected and weighed.
10. In all cases, sap flow is expressed in g H_2O(exudate) g^{-1} root dry weight (DWR) h^{-1} and plotted against pressure (MPa), with the slope being L value in g H_2O g^{-1} RDW MPa^{-1} h^{-1} (Fig. 6). Flow = g exudate h^{-1}, which represents bulk flow and L = g exudate h $^{-1}$ g RDW MPa^{-1}, which represents root hydraulic conductivity.

3.2 Free Exudation Method

1. For plants grown in hydroponic culture, the nutrient solution of trays is changed and all trays are filled with the same volume (*see* **Note 4**) (Fig. 1b).
2. For plants grown in pots, they are immersed in aerated nutrient solution; resembling hydroponic conditions (*see* **Note 5**).

Fig. 7 Free exudate method (**a**) example of silicon tube adapted to the stem, (**b**) the silicon tubes are closed with empty Eppendorf tubes to avoid evaporation of the exudate

3. Aerial parts are excised with a razor blade just below the cotyledons or their stems are cut just below the first branches (*see* **Note 1**) (Fig. 5a).
4. A pipette tip connected to a silicon tube is attached to the stem (*see* **Note 2**) (Fig. 7a).
5. The liquid exuded from the root in the first 15 min is discarded to avoid phloem contaminations.
6. Plants are maintained under exuding conditions for 90 min, the silicon tubes might be closed with empty Eppendorf tubes (Fig. 7b) to avoid exudate evaporation (*see* **Note 6**).
7. The exudates are collected in a new Eppendorf tube (previously weighed) that is weighed to calculate total exudate (*see* **Note 7**).
8. Root dry weight of each plant is determined after oven-drying for 2 days at 75 °C.
9. Osmolality of exuded sap and solution of the trays is determined using a cryoscopy osmometer (*see* **Note 8**) (Fig. 3c).
10. The flow as $Jv = g\ H_2O\ g^{-1}$ RDW h^{-1} is expressed in a root dry weight basis where g H_2O is the amount of exudate, DWR is root dry weight.
11. Osmotic root hydraulic conductance is calculated as $L = Jv/\sigma \times \Delta\Psi$, where Jv is exuded sap flow rate, σ the osmotic reflection coefficient and $\Delta\Psi$ is the osmotic potential gradient between the exuded sap and the solution. According to Fiscus (1986), we assume that reflection coefficient (σ) is close to 1 (0.98 ± 0.01), so we did not include it in the calculation of L.

3.3 High-Pressure Flow Meter Method (HPFM)

Root hydraulic conductivity (L) is calculated using a high-pressure flow meter (Fig. 2) (Lt, kg cm^{-3} s^{-1} MPa^{-1}) (HPFM, Dynamax, Inc., Houston). User manual and software instructions are freely available at: http://www.dynamax.com/products/plant-hydraulic-conductance/hpfm-generation-3-high-pressure-flow-meter.

1. Make the connection between compressed air tank, HPFM, and PC.
2. Open the software (HPFM3) and open a new folder.
3. Before starting the measurements, it is advisable to carry out the Zero Flow Volts to check the calibration and precision of the machine.
4. Select the plant and cut the stem. Use pressure couplings in function of the stem diameter.
5. There are several flows (the flows are shown in Fig. 2 like tubes of different colors that come out of two cylinders) through which water can enter the plant. Then, you have to select the one that best suits the species that is being analyzed. In case you select an incorrect flow, the measurements will be over-scale.
6. Pressurized water is injected into the root from 0 to 0.5 MPa in the transient mode. The transient mode makes measurements of the flow each few seconds (important for plants, which have low conductance and use orange and green tubes to measure Kr).
7. The program (HPFM3 software) collects the data obtained and makes the regression to calculate the hydraulic conductance (Kr).
8. L is determined by dividing Kr by the root dry weight. In case of no access to the whole root system, indirect determinations like analyzing the root diameter and Kr of part of the root must be done [8].

4 Notes

1. Cutting the stem to eliminate the aerial part should be done with a sharp blade, making a clean cut so as not to obstruct any xylem vessel. The use of scissors or reused blades is not appropriate.
2. Constantly check that the pipette tip or the tube is properly adjusted to the stem of the plant. Sometimes, small drops are seen that come out of the tube and run down the stem, producing loss in the amount of exudate collected. In this case, it is recommended to collect exudate that is available to date and replace the plastic tube. If necessary, another piece of the stem (0.5 cm) might be cut with a sharp blade before putting the

new tube, since sometimes the xylem vessels may become clogged due to handling.

3. The pressures to which the plant is subjected and the time taken to collect the exudates will vary according to the type of plant and the treatment to which they are subjected. It is possible to use different pressures between 0.1–0.6 MPa and time of exudates collection between 1 and 10 min.
4. In plants grown in hydroponic culture, we recommend changing the nutrient solution. In case of growing several plants per tray, it is recommended to separate the roots from each other (comb them) to prevent the roots from being broken at the time of taking the measurements, since some plants will be frozen for molecular studies and others for quantifying L.
5. Check that air pumps are bubbling correctly and have similar air flow intensities. This is very important when plants come from hydroponic cultivation.
6. In plants grown in pots with soil and subjected to water stress, extend the time of exudation, since they need more time to collect the exudate required to perform the quantification.
7. The Eppendorf tubes used to collect total exudates will be weighed empty and after being filled, the difference will be total exudates collected.
8. In case of plants that exude very little, dilutions can be made to measure their osmotic concentration.

Acknowledgment

This work was supported by AGL-2017-88341-R grant from the Ministry of Science, Innovation and Universities of Spanish Government.

References

1. Calvo-Polanco M, Sanchez-Romera B, Aroca R (2014) Mild salt stress conditions induce different responses in root hydraulic conductivity of phaseolus vulgaris over-time. PLoS One 9(3): e90631
2. Sánchez-Romera B, Calvo-Polanco M, Ruiz Lozano JM, Zamarreño AM, Arbona V, García-Mina JM, Gómez- Cadenas A, Aroca R (2018) Involvement of the def-1 mutation in the response of tomato plants to arbuscular mycorrhizal simbiosis under well-watered and drought conditions. Plant Cell Physiol 59(2):248–261
3. Li GW, Tillard P, Gojon A, Maurel C (2016) Dual regulation of root hydraulic conductivity and plasma membrane aquaporins by plant nitrate accumulation and high-affinity nitrate transporter NRT2.1. Plant Cell Physiol 57 (4):733–742
4. Hassini I, Martinez-Ballesta MC, Boughanmi N, Moreno DA, Carvajal M (2017) Improvement of broccoli sprouts (Brassica oleracea L. var. italica) growth and quality by KCl seed priming and methyl jasmonate under salinity stress. SCI HORT 226:141–151
5. Fiscus EL (1986) Diurnal changes in volume and solute transport coefficients of *Phaseolus* roots. Plant Physiol 80:752–759

6. Steudle E, Peterson CA (1998) How does water get through roots? J Exp Bot 49:775–778
7. TyreeMT YS, Cruiziat P, Sinclair B (1994) Novel methods of measuring hydraulic conductivity of tree root systems and interpretation using AMAIZED. Pant Physiol 104:189–199
8. Tyree MT, Patino S, Bennink J, Alexander J (1995) Dynamic measurements of root hydraulic conductance using a high- pressure flowmeter in the laboratory and field. J Exp Bot 46:83–94

Part II

Biotic Interactions

Chapter 4

Role of Jasmonates in Beneficial Microbe–Root Interactions

Veronica Basso and Claire Veneault-Fourrey

Abstract

The phytohormone jasmonate (JA) modulates various defense and developmental responses of plants, and is implied in the integration of multiple environmental signals. Given its centrality in regulating plant physiology according to external stimuli, JA influences the establishment of interactions between plant roots and beneficial bacteria or fungi. In many cases, moderate JA signaling promotes the onset of mutualism, while massive JA signaling inhibits it. The output also depends on the compatibility between microbe and host plant and on nutritional or environmental cues. Also, JA biosynthesis and perception participate in the systemic regulation of mutualistic interactions and in microbe-induced resistance to biotic and abiotic stress. Here, we review our current knowledge of the role of JA biosynthesis, signaling, and responses during mutualistic root–microbe interactions.

Key words Jasmonate, Mutualism, Root–microbe interactions, Mycorrhiza, Root–nodule symbiosis, Autoregulation, Induced-systemic resistance, Mycorrhiza-induced resistance

1 Introduction

Optimal plant nutrition and resistance to biotic and abiotic stresses result from an interplay between direct plant adaptation and the effects of beneficial plant-interacting microbes. Since root systems absorb water and mineral nutrients, the root-associated microbiome is of particular interest for the fitness of agricultural and silvicultural crops [1, 2]. The root microbiome can improve plant nutrient uptake [3], and resistance to abiotic stress like drought [4] and salinity [5] and belowground and aboveground diseases and pests [6–9].

Understanding the molecular mechanisms of beneficial plant–microbe interaction establishment is pivotal in order to engineer stress-resistant crops [1]. In this respect, simplified models including one host and one microbial species have long been used to investigate beneficial interactions with high ecological relevance, like the arbuscular mycorrhizal (AM) symbiosis [10] and the root–nodule symbiosis (RNS; [11]). Phytohormones play a prominent role in modulating root–microbe interactions as well as in

Antony Champion and Laurent Laplaze (eds.), *Jasmonate in Plant Biology: Methods and Protocols*, Methods in Molecular Biology, vol. 2085, https://doi.org/10.1007/978-1-0716-0142-6_4,

determining the development of symbiotic structures [11–13]. They are also major regulators of the growth–defense balance. Plant defense is mainly orchestrated by a molecular crosstalk between jasmonate (JA), salicylate (SA), and ethylene (ET) signaling [14]. SA and JA signaling mainly act antagonistically, inhibiting each other [14], and some pathogens hijack plant defense by manipulating the JA–SA crosstalk [15, 16]. Since some symbiotic microbes have common ancestors with pathogens or saprotrophs [13, 17], it is interesting to investigate if tuning of host JA-mediated defense allows colonization by beneficial microbes, while blocking excess invasion [18]. Also, jasmonic acid and its derivatives, collectively known as jasmonates (JAs), are emerging as regulatory hubs of the plant development-to-defense trade-off [19, 20]; therefore, they likely integrate developmental and environmental signals in order to modulate interactions whose impact on plant fitness depends on genetic or environmental factors [21]. Conversely, beneficial microbe-induced systemic resistance (ISR) to pathogens often implies enhanced sensitivity to JA/ET signaling [22], indicating multifaceted roles of JA in beneficial interactions. Indeed, JA is involved in both establishment and function of mutualistic interactions, even if contradictory findings have been reported. This review summarizes current knowledge on the diversified roles of JA in microbial recruitment, symbiosis development, and physiological consequences on host plants throughout different types of root–microbe beneficial associations.

2 JA in Arbuscular Mycorrhiza Development

Arbuscular mycorrhiza is mutualistic association taking place between the roots of ~80% of land plants and some classes of fungi of the *Glomeromycotina* sub-phylum. Arbuscular mycorrhizal fungi (AMF) can increase plant uptake of water, nitrogen (N), phosphorus (P), and other mineral elements, in exchange for sugar and lipids [23], improving plant nutrition and health [24]. Root infection by AMF begins with the formation of a hyphopodium at the root surface, followed by constitution of a pre-penetration apparatus (PPA) and hyphal intracellular penetration through epidermal and cortical cell layers. Eventually, AMF spread longitudinally through the root cortex and generate ramified structure called arbuscules that are sites of nutrient exchanges between the symbiotic partners [10].

2.1 JA Biosynthesis and Signaling During AM Development

JA biosynthesis, secretion, and associated responses modulate several stages of AM symbiosis development, with different outputs depending on phylogenetic, genetic, and environmental factors.

First, secretion of JA or of JA-induced compounds may trigger fungal hyphae elongation for a quicker contact to the host root.

In fact, root exudates of *Lotus japonicus* grown in high Red/Far Red (R/FR) light conditions, mimicking optimal light, as well as 0.5 μM exogenous JA, stimulated hyphal elongation of *Rhizophagus irregularis* [25]. However, since R/FR light signaling represses JA signaling as part of the shade-avoidance syndrome (SAS; [26]), it remains to be understood if light cues influence JA secretion in the early phase of AM symbiosis and/or if they participate to the systemic regulation of mycorrhization at the mature stage.

No JA biosynthesis was reported during early stages of AM development, that is, hyphopodium or prepenetration apparatus (PPA) formation. However, *Gigaspora margarita* inoculation of *Medicago truncatula* roots induced the expression of some JA-responsive genes during PPA formation [27]. JA signaling at this stage might counteract the effect of transient SA accumulation upon first contact between the two symbiotic partners [28–30].

On the other hand, JA biosynthesis and signaling pathways are active in late stages of AM colonization in several plant species [31–35]. JA accumulates in arbuscule-containing cells of barley [31] and *M. truncatula* [34]. Local JA production in the roots has a positive effect on mycorrhization, as demonstrated by reduced colonization and/or arbuscule frequency in mutant *M. truncatula* and tomato roots impaired in JA biosynthesis [34–36]. Conversely, constitutive high levels of JA resulted in enhanced arbuscule frequency in tomato [36], but slightly reduced it in *M. truncatula* [37], suggesting the involvement of regulatory feedback mechanisms in some plant species. Interestingly, the positive effect of JA signaling in AM symbiosis is independent of the canonical JA-Ile receptor COI1 in tomato. Tomato *jai1-1* mutants, bereft of COI1, are more mycorrhized than wild-type tomato plants [35, 38]. Therefore, JA signaling might negatively influence arbuscule development through the COI1-dependent signaling pathway and positively influence it through an unknown, independent pathway [30]. Also, it is still controversial whether JA biosynthesis is needed locally or systemically for promoting its action on AM symbiosis. Moderate and systemic activation of JA signaling may also promote AM, since low dosage or sporadic application of JA/MeJA on leaves induced higher colonization and/or arbuscule frequency in garlic, tomato, cucumber, and *M. truncatula* [36, 39–41]. On the other hand, massive or repeated JA/MeJA application to leaves or roots in several plant species reduced AM colonization with concomitant induction of plant defenses and root growth arrest [38, 40, 42, 43].

Overall, the role of JA in AMF colonization depends on plant and fungal genotypes and on environmental conditions. In fact, JA biosynthesis and/or signaling are necessary neither for the establishment of the semi-parasitical interaction between *Nicotiana attenuata* and *R. irregularis* [44] nor for the AM symbiosis in rice [42], indicating the dispensability of JA-mediated regulation for some fungus–host plant pairs. Moreover, JAs differentially accumulated in soybean, tomato, and maize inoculated with either *R. irregularis* or *Funneliformis mosseae*. Indeed, colonization by the less successful fungus *F. mosseae* promoted oxylipin biosynthesis and signaling to a greater extent than by *R. irregularis* [45]. This suggests that JA accumulation suppresses colonization by poorly compatible AMF, possibly through induced-systemic defense. Interestingly, exogenous JA-triggered promotion or delay of AM symbiosis depends on the P_i status of cucumber plants [40]. Taken together, these results suggest that sustained JA signaling restricts AM colonization according to the fungus–host compatibility and to environmental conditions, as part of a defense-to-development trade-off.

2.2 Possible Functions of JA Signaling in Mature AM

Several possible functions of JA signaling in mature mycorrhizae are emerging. We propose three non-mutually exclusive hypotheses supported by several lines of evidence.

First, JA signaling might help reshape cortical cells for arbuscule accommodation. JA can, indeed, induce microtubule depolymerization [46] and massive cytoskeletal remodeling occurs during arbuscule development [47, 48]. Moreover, the *M. truncatula β-TUBULIN* gene is upregulated both during AM symbiosis [49] and in response to JA treatment [37]. Also, JAs might modulate cell wall composition and vacuole osmotic pressure, similarly to their action during tuber development [50].

Second, JA accumulation in arbuscule-containing cells possibly enhances the carbon sink strength of arbuscules in order to sustain symbiotic nutrient exchanges. This hypothesis is supported by several pieces of evidence: (1) low JA concentrations sprayed on a single leaf privilege carbon partitioning toward roots in *Populus tremuloides*, *Populus nigra* [51] and tomato [52]; (2) the JA biosynthetic genes *AOS* and *AOC* are induced by glucose in tomato flowers and barley seedlings [53, 54]; (3) AMF inoculation induces carbon accumulation in tomato roots [55]; (4) JA-responsive genes are simultaneously activated in arbusculated cells and in phloem companion cells in mycorrhizal barley, suggesting their involvement in nutrient transport [31]; (5) the expression of the cell wall invertase gene *Lin6* is induced both by AM inoculation [56] and JA treatment [57]; (6) the structure and proteome of plastids of arbusculated *M. truncatula* cells display a nutrient starvation signature: therefore, JA biosynthesis might be induced to increase shoot-to-root carbon flux in consequence of amyloplastic stress [58]. We propose that if the main function of JA signaling in arbusculated cells was to enhance carbon sink strength, overexpressing genes involved in symbiotic sugar transport and conversion in a JA-deficient background would compensate the reduced arbuscule frequency phenotype.

Finally, JA signaling may inhibit excessive *in planta* fungal colonization. This hypothesis is supported by the fact that treatment with a high dose of JA/MeJA reduces or delays AM colonization [38, 42, 43] and that JA-insensitive tomato mutants are more mycorrhized than wild-type plants [38]. JA signaling can lead to enhanced production of defense flavonoids in arbuscule-containing cells [59]. In addition, the 9-LOX biosynthetic pathway, producing a range of antimicrobial oxylipins [60] is activated in mycorrhizal tomato roots [61], partially depending on functional JA biosynthesis and signaling [35]. Alternatively, AM fungi may induce flavonoids or oxylipin production to prime the host plant for defense against pathogenic microbes (*see* Subheading 6.2).

In conclusion, JA biosynthesis and signaling may serve different purposes across the various developmental stages of AM symbiosis. In many plant species, moderate local or systemic JA accumulation promotes arbuscule formation. However, phylogenetic, genetic, and environmental factors strongly influence the output of activated JA signaling on the extent of AM symbiosis, up to the point that sustained JA response counteracts mycorrhizal colonization. JA accumulation in arbuscule-containing cells may mediate cell reshaping, symbiotic nutrient exchanges, or regulation of excessive *in planta* fungal colonization. Systematic studies on the genetic elements controlling JA production, conversion, signaling, and downstream gene regulation will be necessary to draw a clear model of the role of JAs in the development of AM symbiosis (Fig. 1).

3 JA in Root–Nodule Symbiosis Development

There are two types of root–nodule symbioses (RNS), the rhizobia–legume symbiosis and the actinorhizal symbiosis. Rhizobia–legume symbiosis is the mutually beneficial interaction between several species of the *Rhizobia* paraphyletic group of α- and β-proteobacteria and the roots of most legumes. The actinorhizal symbiosis results from the interaction between some members of the Cucurbitales, Fagales, and Rosales orders and N-fixing actinorhizal bacteria of the *Frankia* genus. In both cases, the outcome of the interaction is the generation of a novel organ, the N-fixing nodule, in which bacteria differentiate to convert atmospheric N into an organic form, available for the plant [62]. Given the lack of evidence of any role of JA in the establishment of actinorhizal symbiosis, in the following paragraphs, we will focus on JA-mediated modulation of the rhizobia–legume symbiosis.

3.1 JA Biosynthesis and Signaling During RNS

As in the case of AM symbiosis, there are reports of a positive [63], negative [64–66], and neutral [41] role of JA biosynthesis and signaling in RNS.

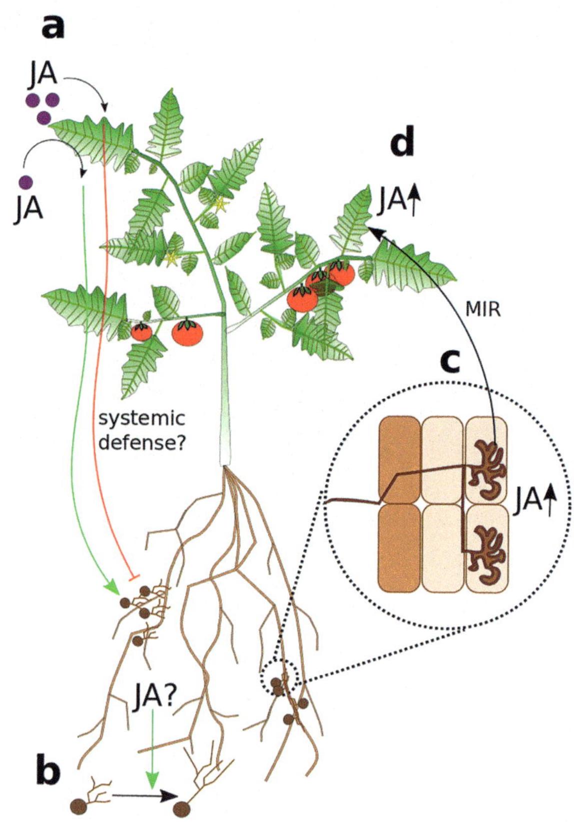

Fig. 1 The role of JAs in development of AM symbiosis. (**a**) High doses of exogenous JAs in leaves inhibit the establishment of AM symbiosis, possibly through activated systemic resistance; however, low doses promote it, through unknown mechanisms. (**b**) JA may promote fungal hyphae elongation, thereby facilitating hyphae–root contact. (**c**) The level of JA rises in arbusculated cells of model plant species, from tomato to *M. truncatula*. In the same species, JA biosynthesis or JA-mediated responses are important for the occurrence of arbuscules at high frequency. JA accumulation in arbuscule-containing cells may help (1) reshape the cell for arbuscule accommodation; (2) induce a carbon sink status, reinforcing the flow of nutrients from shoot to arbuscules; (3) suppress excessive fungal invasion. (**d**) Some late-stage mycorrhizal plants increase their shoot JA level. This possibly mediates mycorrhiza-induced resistance (MIR) to abiotic and biotic stress

In legumes, JA regulates the early stages of bacterial colonization, that is, root hair curling and infection thread (IT) formation, by influencing the plant–bacteria communication mechanism represented by bacterial Nod factor-triggered Ca^{2+}-spiking in root cortical cells [67]. Exogenous JA treatment stimulated *Nod* gene expression and Nod factor production by isolated rhizobia [68, 69]. Also, very moderate JA application (0.1 μM) to growth medium slightly increased IT and nodule number in *L. japonicus* [63]. Therefore, secreted JA or JA-induced flavonoids may stimulate rhizobial expression of *Nod* genes and chemoattraction to the host root [70]. However, increasing doses of exogenous MeJA inhibited root curling, IT formation and nodule primordia organogenesis of *L. japonicus* in a concentration-dependent manner [64] and progressively decreased *M. truncatula* sensitivity to Nod factor [65]. Conversely, 1 h treatment with Nod factor decreased the *M. truncatula* root content of oxylipin precursors [71]. Altogether, these results suggest that active JA signaling in the precontact phases of RNS helps to activate the symbiotic molecular dialogue and that subsequently, upon Nod factor signaling, reduced oxylipin biosynthesis allows bacterial infection. This hypothesis is supported by the finding that the *L. japonicus* JA biosynthetic genes *AOC* and *12-OPDA reductase* were strongly upregulated at −4 days post rhizobial inoculation, to almost revert to uninfected control expression levels later on [72]. We thus propose that, similarly to AM symbiosis, moderate and transient JA signaling promotes RNS, while massive or sustained JA signaling dampens RNS by influencing the success of early colonization events.

In distinct N-fixing bacteria–plant interactions, oxylipin biosynthesis is enhanced in parenchymatous cells of mature nodules [73–77], although this phenomenon may be transient or species-specific [72, 78]. Since bioactive JA-Ile levels do not vary upon nodulation [76, 79] and neither impaired nor enhanced root JA biosynthesis does impact RNS in *M. truncatula* [79], the production of other oxylipins type may play a more significant role in nodulation. However, JA signaling seems to negatively affect bacteroid performance, since *Astragalus sinicus* hairy roots silencing the *JAZ1* gene, thus constitutively activating JA signaling [80], displayed reduced bacteroid differentiation and subsequent decrease in nitrogenase activity [81]. AsJAZ1 interacts with the leghemoglobin AsB2510 [81]. Leghemoglobins generate the microaerobic environment necessary for nitrogenase activity [82] and are fundamental for RNS [83]. Therefore, JA signaling may modulate symbiotic N fixation through tuning of leghemoglobin activity. The contradiction between absence of phenotype during constitutive JA biosynthesis [79] and severe nodulation phenotype when JA signaling is constitutively activated [81] hints at a modified JA sensitivity in nodules and needs further investigation.

3.2 JA Signaling in the Autoregulation of Nodulation (AUT)

JA signaling contributes to the autoregulation of nodulation (AUT), a pathway suppressing nodule formation upon already established RNS [84]. The fundamental genetic component of the AUT pathway is the LjHAR1/MtSUNN/GmNARK CLAVATA1-like leucine-rich repeat (LRR) receptor kinase [85–88], acting in the shoot to perceive a nodule-derived signal mediated by CLE peptides [89, 90]. Subsequently, LjHAR1/MtSUNN/GmNARK codifies still partially unknown signals that suppress nodulation systemically. The corresponding mutants, *L.*

japonicus har1, *M. truncatula sunn* and soybean *nark*, are hypernodulating [85–88]. Non-inoculated *nark/sunn* mutants show high JA levels and expression of JA biosynthetic/response genes [66, 91], whereas the same genes are downregulated in the leaves of nodulating wild-type *M. truncatula* [91]. This indicates that LjHAR1/MtSUNN/GmNARK-dependent systemic dampening of JA biosynthesis/signaling may restrict nodulation. Since decreased accumulation of the miR2111 microRNA has been proposed to be the AUT shoot-derived signal in *L. japonicus* [92], it is worth verifying if the expression of the *MIR2111* genes is JA-dependent. However, additional AUT mechanisms must exist, since shoot treatment with the JA biosynthesis inhibitor (n-propyl gallate) inhibited nodulation in the hypernodulating *nark* mutant, but failed to suppress it to wild-type levels [91]. Moreover, optimal JA homeostasis in shoots is crucial for RNS: in fact, both impaired JA biosynthesis and exogenous JA treatment reduce nodule number, both in wild-type and hypernodulating AUT mutants [64, 66, 91, 93]. Moderate JA signaling could counteract AUT through inhibited auxin synthesis and transport [94]. Indeed, hypernodulating *nark* mutant roots failed to increase their auxin content upon rhizobial inoculation [95]; and long-distance auxin transport is inhibited in *sunn* mutants [96], suggesting that auxin accumulation in roots inhibits excessive nodulation. To integrate all these results, we propose that: (1) moderate or transient JA systemic signaling, as in non-nodulating plants, alters auxin gradients to promote nodulation; (2) very low levels of foliar JA, induced by action of LjHAR1/MtSUNN/GmNARK, fail to restrain auxin accumulation in roots, with a consequent inhibition of nodulation; (3) massive and sustained JA signaling in the shoot, as obtained through exogenous application, induces a systemic immune response preventing rhizobial colonization. The latter assumption, however, contrasts with the observation that attack by the foliar chewing insect *Spodoptera exigua*, which is supposed to trigger JA-dependent defenses, increases nodulation of *M. truncatula* plants [97]. Nevertheless, JA-mediated responses to herbivore attack result from the integration of multiple environmental signals, for example, light availability [19]. Similarly, phytochrome signaling probably dampens JA signaling in order to repress RNS, as suggested by decreased nodulation, JA-Ile levels and/or JA marker gene expression in shade-grown wild-type *L. japonicus* and in light-insensitive *phyB* roots [63]. The interaction between LjHAR1/MtSUNN/GmNARK-mediated and PHYB-mediated inhibition of nodulation has not yet been investigated. We propose the restriction of JA biosynthesis to be downstream of both pathways. This hypothesis can be verified through phenotyping of hypernodulating *phyB* and JA-deficient mutant legumes and their crossings.

In conclusion, moderate JA accumulation may promote the RNS molecular dialogue, while excessive JA accumulation inhibits RNS, probably through enhanced systemic defense. JA signaling, but not JA biosynthesis, might regulate symbiotic N fixation. Finally, the inhibition of JA biosynthesis, by either the AUT pathway or light signaling, contributes to the systemic restriction of nodulation, possibly through altered auxin homeostasis. Therefore, the correct balance of shoot oxylipins is crucial for optimal RNS (Fig. 2).

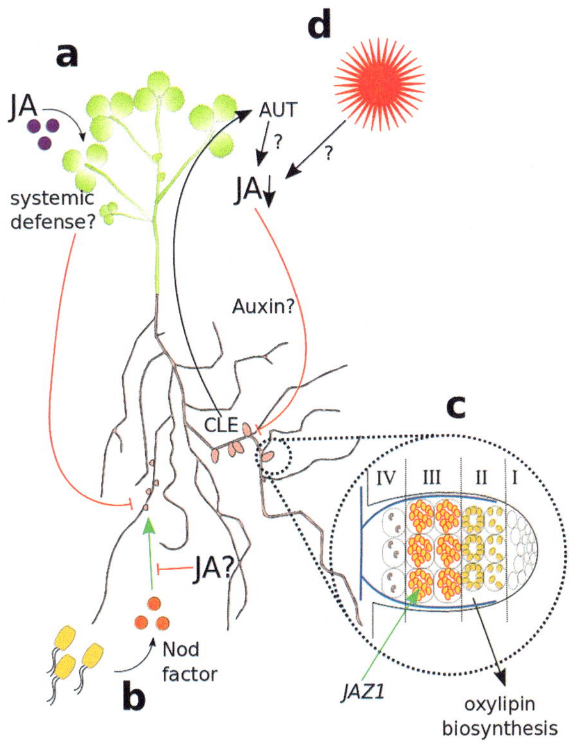

Fig. 2 The role of JAs in RNS. (**a**) Exogenous JA treatment of leaves reduces the number of bacterial IT, nodule primordia and mature nodules in several plant species, possibly through enhanced systemic defense. (**b**) JA may promote the production of rhizobial Nod factor, but impair its perception, thus counteracting the formation of nodule primordia. (**c**) Oxylipin biosynthetic enzymes are not expressed in N-fixing cells and repression of JA signaling through JAZ1 is necessary for bacteroid differentiation and nitrogenase activity: this suggests a negative impact of JA signaling in bacteroid development and N fixation. (**d**) AUT, triggered by CLE peptides, and SAS, triggered by low R/FR radiation, may systemically inhibit JA biosynthesis/signaling, thus releasing the repression over auxin biosynthesis and transport toward the roots; in turn, root auxin accumulation may repress RNS

4 JA in Ectomycorrhizal Symbiosis Development

The ectomycorrhizal symbiosis (ECM) is a mutually beneficial interaction occurring between the roots of about 6000 tree species and more than 20,000 species of fungi of the *Basidiomycota* and *Ascomycota* phyla [13]. ECM fungi are crucial for forest ecosystems, since (1) they enhance tree mineral nutrition [98, 99], (2) they provide protection against biotic and abiotic stress [100] and (3) they participate in the carbon and nitrogen cycles [101, 102]. Fungal hyphae form a mantle around lateral roots and subsequently penetrate through the apoplastic space between epidermal and cortical cells, to generate a ramified structure called the Hartig net, the probable site of nutrient exchange between the symbiotic partners [103]. The molecular determinants of ECM development are much less known with respect to other intimate kinds of plant–microbe symbioses, and may vary considerably, since ECM evolved multiple times [104].

In the model system constituted by trees of the *Populus* genus and by the Basidiomycete *Laccaria bicolor*, suppression of JA signaling is pivotal for the establishment of the symbiosis. In fact, the effector Mycorrhiza-Induced Small Secreted Protein 7 (MiSSP7) from *L. bicolor*, secreted upon initiation of the symbiotic program [105], penetrates the nuclei of poplar root cortical cells and interacts with the poplar co-receptor of jasmonic acid JAZ6, stabilizing its repression over JA signaling [105, 106]. Since both inoculation with *missp7* silenced strains of *L. bicolor* and very low concentrations of exogenous MeJA impaired Hartig net formation [105, 107], MiSSP7-mediated inhibition of the JA signaling pathway is necessary for ECM development.

However, JA signaling may have distinct functions at different mycorrhizal stages. At early mycorrhizal stage (1 week post inoculation), JA signaling may promote moderate immunity (Basso et al., in preparation). Furthermore, JA-downregulated genes involved in biosynthesis of and/or response to auxin, gibberellin, ethylene, cytokinin, strigolactone and brassinosteroid are induced in early ECM (Basso et al., in preparation), suggesting suppression of the crosstalk between JA and other phytohormone signaling pathways during ECM development. In mature ECM (2 weeks post inoculation), JA signaling may play a role in root cell reshaping; necessary for Hartig net formation (Basso et al., in preparation). Also, it has been suggested that active JA signaling triggers plant immunity, in order to restrict fungal growth within the root at late stages [107]. The use of poplar mutants defective in different branches of JA signaling will possibly highlight a multifaceted role of JA during ECM development.

5 JA in Other Beneficial Rhizospheric/Endophytic Interactions

In the era of omics technology, it has become easier and easier to tackle complex systems. It is thus possible to expand the field of plant–microbe interactions from the study of single host–single microbe systems to investigate rhizospheric community dynamics. Indeed, N-fixing rhizobia and mycorrhizal fungi are just part of a highly diversified group of root-associated beneficial microbes called plant-growth promoting rhizobacteria (PGPR) and fungi (PGPF). Such microbes are attracted toward the rhizosphere, that is, the area of soil occupied by roots and by plant exudates [108] and can become epiphytes colonizing the root surface, or endophytes populating inner root tissues [109]. These microbes are a hidden reservoir of useful metabolic functions for plant nutrition and health [110, 111] and are being widely exploited as biocontrol agents in agriculture [112, 113]. However, the output of such interactions on plant fitness depends on plant genotype [114, 115] and environment [55] and on the type of stress to which the host plant is subjected [116, 117]. Therefore, to achieve optimal fitness, plants must integrate developmental and environmental cues prior to regulation of rhizospheric or endophytic interactions. JA signaling, the master switch between development and defense [19, 20], may contribute to both processes.

First, secreted JA or JA-triggered secondary metabolites may remodel composition, diversity, and richness of microbial communities [108]. For example, various diseases due to necrotrophic plant pathogens induce the development of disease-suppressive soils [118–120], probably through increased abundance of particular microbial species upon plant-derived rhizospheric cues; the latter are possibly part of the JA-mediated defense against necrotrophic pathogens [14]. In support of this hypothesis, Arabidopsis leaf treatment with low concentrations of MeJA increased rhizospheric abundance of bacteria closely related to pest-protective species [121]. The same treatment in wheat provoked a shift in the endophytic bacterial community [122]. In both cases, the consequences of microbial recruiting on plant fitness need to be investigated. In addition, JA perception seems to control the composition of root-associated microbial communities. Indeed, roots of JA-insensitive Arabidopsis mutants reduced secretion of asparagine, tryptophan, glutamic acid, ornithine, and galactinol compounds implied in the chemotaxis of several beneficial bacterial species [123]. These results suggest that JA-mediated production of secondary metabolites, rather than direct JA secretion, reshapes root-associated microbial communities.

Second, JA signaling may determine the level of compatibility between beneficial microbes and hosts, as suggested in case of mycorrhizal fungi [45] and N-fixing *Azoarcus* spp. bacteria

[124]. On the other hand, induction of JA signaling, counteracting SA signaling, is advantageous for some beneficial microbes. For example, shoot MeJA treatment promoted the colonization of Arabidopsis roots by the disease-protective endophyte *Trichoderma harzianum* T-78 [125]. Also, the broad-host-range basidiomycete *Piriformospora indica*, conferring increased yield, abiotic stress, and disease resistance to plants [126, 127], must escape SA-mediated hypersensitive response in the first, biotrophic phase of its root colonization and achieve so through induction of JA signaling [128], possibly through small secreted proteins [129]. However, JA signaling seems to inhibit the subsequent, necrotrophic phase of colonization, since several Arabidopsis mutants impaired in JA biosynthesis and signaling were more colonized by *P. indica* at a later stage [128].

In conclusion, evidence suggests that endogenous JA production and its associated responses are implied in reshaping root-associated microbial communities in various plant species, especially in case of pathogen attack, with a putative protective outcome. Moreover, JA signaling may finely tune the level of colonization by beneficial microbes according to host–microbe compatibility or environmental cues.

6 JA in Beneficial Microbe-Triggered Systemic Resistance

Since the first discoveries of disease-suppressive soils [130], it was speculated that plants under pathogen attack may attract root microorganisms capable of priming host defenses for better performance in case of subsequent biotic stress. Priming is the main mechanism of induced systemic resistance (ISR) by plant growth–promoting rhizobacteria (PGPR) or fungi (PGPF) (reviewed in [111, 131, 132]). A similar phenomenon, triggered AMF, is mycorrhiza-induced systemic resistance (MIR).

6.1 Induced Systemic Resistance (ISR) by Plant Growth–Promoting Rhizobacteria (PGPR) and Fungi (PGPF)

Priming by ISR usually does not involve a direct induction of defense gene expression [132, 133] or alteration of hormonal levels [116, 134] upon presence of the beneficial microbe, even if, upon inoculation with some PGPFs, shoot JA accumulation or enhanced expression of JA biosynthetic genes have been reported [135, 136]. In several plant-beneficial microbe pairs, ISR increases sensitivity to JA and ET upon subsequent pathogen challenge, enhancing defense mechanisms controlled by the JA/ET signaling pathway [22, 132]. As a proof of principle, ISR is mostly effective against JA-sensitive chewing insects, rather than against SA-sensitive sap-sucking insects [117, 137, 138], even though the range of bacteria, oomycetes, fungi, and viruses against which ISR is effective exceeds the variety of organisms usually contrasted through JA signaling [111, 139]. For many PGPR and PGPF genetic evidence in Arabidopsis, tomato, cucumber, and rice showed that JA and/or ET signaling, as well as the master switch of the SA–JA crosstalk, *NPR1*, are necessary for ISR [116, 132, 134, 140–152]. In Arabidopsis, the root-specific transcription factor MYB72 is a central node for early ISR signaling upon *Pseudomonas simiae* WCS417r, *P. putida* WCS358r or *Trichoderma asperellum* T34 inoculation [153, 154]. MYB72 interacts with the ethylene-responsive transcription factor EIL3 and acts upstream of ET and JA signaling [153]. Therefore, a possible model of ISR signaling depicts activity of EIL3-MYB72 at the contact site with the beneficial microbe, followed by NPR1-dependent systemic priming of JA/ET-induced defenses [111, 139]. However, the nature of the locally perceived signal, of the systemic priming signal and of the molecular mechanism of priming is still elusive.

In addition to the widespread JA/ET signaling priming, some beneficial rhizobacteria prime their host for enhanced SA-dependent defenses [155–157], while others, as well as several PGPFs of the *Trichoderma* genus, are able to prime both the SA and the JA/ET pathway [125, 158–161]. Other PGPRs or PGPFs, such as the endophytic fungus P. indica, prime plant defense through JA/ET- and SA-independent mechanisms [127]. Given the wide variety of priming mechanisms, inoculation with multiple beneficial microbes might enhance plant resistance in consequence of the activation of multiple pathways. For example, in cucumber, co-inoculation of PGPF *T. harzianum* and PGPR *Pseudomonas* sp. enhanced resistance to the stem pathogen *Fusarium oxysporum* by activating both JA- and SA-dependent defense responses [162]. However, the molecular mechanism of the JA–SA crosstalk–induced systemic resistance has not been investigated yet.

6.2 Mycorrhiza-Induced Resistance (MIR)

A similar phenomenon to ISR is mycorrhiza-induced resistance (MIR) (reviewed in [163–165]). Mycorrhized plants are more resistant to a wide range of soil-borne pathogens [166], while the impact on resistance to aboveground attackers is highly variable. In general, biotrophic pathogens and pierce-sucking insects perform better on leaves of mycorrhizal plants, even if increased tolerance was observed in terms of plant mass and yield [167]. Conversely, AM promotes resistance against necrotrophic pathogens [168–170] and leaf-chewing insects [171]. One leading mechanism driving MIR is systemic priming of plant defenses, occurring both above and belowground [168, 172–176]. Priming leads, upon pathogen challenge, to a more rapid and effective induction of defense responses, like expression of pathogenesis-related genes and callose deposition [168, 174, 175, 177]. However, MIR is more effective in specific host–AMF pairs [172, 178–181].

The role of JA signaling in MIR is supported by several pieces of evidence: (1) AM confers protection against necrotrophic pathogens and leaf-chewing insects, the immunity to which is mediated by JA [14]; (2) MIR-enhanced defense includes JA-dependent genes (Pozo et al. [182]); (3) at late stages of AM colonization, JA levels increase in leaves of *M. truncatula* (Isayenkov et al. [34]) and tomato (Nair et al., [176]); (4) AM colonization activates JA biosynthesis and responses in leaves of *M. truncatula* [55, 182, 183]; (5) intact JA biosynthesis is required for MIR in tomato [175, 176]; (6) similarly to ISR, priming by MIR involves also enhanced JA sensitivity, since shoots of mycorrhizal tomato plants highly expressed *COI1* [176] and responded earlier to foliar JA application [182]; and (7) tomato susceptibility to *Botrytis cinerea* was inversely correlated with the production of oxylipins and flavonoids in mycorrhizal, unchallenged plants [55]. However, shoot JA accumulation is not required for MIR in all plant species [136] and depends on the N nutritional status in tomato [55]. JA signaling may contribute to MIR also through partitioning of nutrient resources. In fact, herbivore attack [184] and MeJA application [51, 52] enhance carbon allocation toward roots, possibly helping the plant confront both nutritional and biotic stress [185]. Thus, through priming of defense mechanisms and nutrient allocation, enhanced JA accumulation or sensitivity upon AM colonization might mediate resistance to both biotic and abiotic stress.

7 Conclusion

Jasmonate biosynthesis and signaling play very diversified roles in interactions between root-associated beneficial microbes and plants. A canonical defense phytohormone, JA, often opposes the formation of root–microbe interactions, probably through the activation of systemic defense [38, 40, 42, 43, 64–66, 124]. Indeed, endogenous JA levels seem to inversely correlate with the degree of compatibility between microbes and hosts [45, 124]. To overcome plant immunity, similarly to pathogens, some beneficial organisms hijack JA signaling through the use of effectors to establish symbioses [105, 128]. However, low levels of JA promote AM colonization [36, 39–41] and attract beneficial organisms through modification of root exudates [118–122]. Overall, in many plant species the fine-tuning of JA accumulation and/or signaling determine the colonization success of beneficial microbes.

JA signaling is often involved in the maintenance of deeply connected interactions. For example, enhanced JA signaling in arbuscule-containing cells of some mycorrhizal species may support cell reshaping or fungal carbon nutrition, necessary for the maintenance of AM symbiosis [37]. However, reduced shoot JA accumulation may be important for AUT in nodulating legumes [66, 91]. Nevertheless, sustained JA signaling is deleterious for many interactions. The interplay between JA-mediated stabilization and hampering of beneficial interactions needs to be further investigated. Finally, beneficial organisms can stimulate JA biosynthesis and/or signaling in systemic way in order to prime the host plant for better resistance to biotic and abiotic stress [111, 164, 165].

Since genetic and environmental factors influence the output of JA accumulation and response in mutualistic plant–microbe interactions, JA signaling may integrate different kinds of environmental stimuli and control the energetic defense-to-development trade-off [19, 20]. Further research is necessary to investigate the multifaceted roles of this major regulator of plant physiology and beneficial interactions.

References

1. Lakshmanan V, Selvaraj G, Bais HP (2014) Functional soil microbiome: belowground solutions to an aboveground problem. Plant Physiol 166(2):689–700. https://doi.org/10.1104/pp.114.245811
2. Mercado-Blanco J, Abrantes I, Caracciolo AB, Bevivino A, Ciancio A, Grenni P et al (2018) Belowground microbiota and the health of tree crops. Front Microbiol 9(6):1006. https://doi.org/10.3389/fmicb.2018.01006
3. Tkacz A, Poole P (2015) Role of root microbiota in plant productivity. J Exp Bot 66 (8):2167–2175. https://doi.org/10.1093/jxb/erv157
4. Rolli E, Marasco R, Vigani G, Ettoumi B, Mapelli F, Deangelis ML et al (2015) Improved plant resistance to drought is promoted by the root-associated microbiome as a water stress-dependent trait. Environ Microbiol 17(2):316–331. https://doi.org/10.1111/1462-2920.12439
5. Qin Y, Druzhinina IS, Pan X, Yuan Z (2016) Microbially mediated plant salt tolerance and microbiome-based solutions for saline agriculture. Biotechnol Adv 34(7):1245–1259. https://doi.org/10.1016/j.biotechadv.2016.08.005
6. Berendsen RL, Pieterse CMJ, Bakker PAHM (2012) The rhizosphere microbiome and plant health. Trends Plant Sci 17 (8):478–486. https://doi.org/10.1016/j.tplants.2012.04.001
7. Raaijmakers JM, Mazzola M (2016) Soil immune responses. Science 352 (6292):1392–1393. https://doi.org/10.1126/science.aaf3252
8. Shikano I, Rosa C, Tan C-W, Felton GW (2017) Tritrophic interactions: microbe-mediated plant effects on insect herbivores. Annu Rev Phytopathol 55(1):313–331. https://doi.org/10.1146/annurev-phyto-080516-035319
9. Bakker PAHM, Pieterse CMJ, de Jonge R, Berendsen RL (2018) The soil-borne legacy. Cell 172(6):1178–1180. https://doi.org/10.1016/j.cell.2018.02.024
10. Gutjahr C, Parniske M (2013) Cell and developmental biology of Arbuscular Mycorrhiza Symbiosis. Annu Rev Cell Dev Biol 29 (1):593–617. https://doi.org/10.1146/annurev-cellbio-101512-122413
11. Oldroyd GED, Murray JD, Poole PS, Downie JA (2011) The rules of engagement in the legume-Rhizobial Symbiosis. Annu Rev Genet 45(1):119–144. https://doi.org/10.1146/annurev-genet-110410-132549
12. Gutjahr C (2014) Phytohormone signaling in arbuscular mycorhiza development. Curr Opin Plant Biol 20:26–34. https://doi.org/10.1016/j.pbi.2014.04.003
13. Martin F, Kohler A, Murat C, Veneault-Fourrey C, Hibbett DS (2016) Unearthing the roots of ectomycorrhizal symbioses. Nat Rev Microbiol 14(12):760–773. https://doi.org/10.1038/nrmicro.2016.149
14. Pieterse CMJ, Leon-Reyes A, Van Der Ent S, Van Wees SCM (2009) Networking by small-molecule hormones in plant immunity. Nat Chem Biol 5(5):308–316. https://doi.org/10.1038/nchembio.164
15. Nomura K, Melotto M, He S-Y (2005) Suppression of host defense in compatible plant–pseudomonas syringae interactions. Curr Opin Plant Biol 8(4):361–368. https://doi.org/10.1016/j.pbi.2005.05.005
16. El Oirdi M, El Rahman TA, Rigano L, El Hadrami A, Rodriguez MC, Daayf F et al (2011) Botrytis cinerea manipulates the

antagonistic effects between immune pathways to promote disease development in tomato. Plant Cell 23(6):2405–2421. https://doi.org/10.1105/tpc.111.083394
17. Marchetti M, Capela D, Glew M, Cruveiller S, Chane-Woon-Ming B, Gris C et al (2010) Experimental evolution of a plant pathogen into a legume symbiont. PLoS Biol 8(1): e1000280. https://doi.org/10.1371/journal.pbio.1000280
18. Pieterse CMJ, Van der Does D, Zamioudis C, Leon-Reyes A, Van Wees SCM (2012) Hormonal modulation of plant immunity. Annu Rev Cell Dev Biol 28(1):489–521. https://doi.org/10.1146/annurev-cellbio-092910-154055
19. Kazan K, Manners JM (2011) The interplay between light and jasmonate signalling during defence and development. J Exp Bot 62 (12):4087–4100. https://doi.org/10.1093/jxb/err142
20. Campos ML, Yoshida Y, Major IT, De Oliveira Ferreira D, Weraduwage SM, Froehlich JE et al (2016) Rewiring of jasmonate and phytochrome B signalling uncouples plant growth-defense tradeoffs. Nat Commun 7:12570. https://doi.org/10.1038/ncomms12570
21. Friesen ML, Porter SS, Stark SC, von Wettberg EJ, Sachs JL, Martinez-Romero E (2011) Microbially mediated plant functional traits. Annu Rev Ecol Evol Syst 42(1):23–46. https://doi.org/10.1146/annurev-ecolsys-102710-145039
22. Verhagen BWM, Glazebrook J, Zhu T, Chang H-S, van Loon LC, Pieterse CMJ (2004) The transcriptome of rhizobacteria-induced systemic resistance in Arabidopsis. Mol Plant-Microbe Interact 17(8):895–908. https://doi.org/10.1094/MPMI.2004.17.8.895
23. Wang W, Shi J, Xie Q, Jiang Y, Yu N, Wang E (2017) Nutrient exchange and regulation in Arbuscular Mycorrhizal Symbiosis. Mol Plant 10(9):1147–1158. https://doi.org/10.1016/j.molp.2017.07.012
24. Berruti A, Lumini E, Balestrini R, Bianciotto V (2016) Arbuscular mycorrhizal fungi as natural biofertilizers: let's benefit from past successes. Front Microbiol 6(1):1559. https://doi.org/10.3389/fmicb.2015.01559
25. Nagata M, Yamamoto N, Miyamoto T, Shimomura A, Arima S, Hirsch AM, Suzuki A (2016) Enhanced hyphal growth of arbuscular mycorrhizae by root exudates derived from high R/FR treated Lotus japonicus. Plant Signal Behav 11(6):e1187356. https://doi.org/10.1080/15592324.2016.1187356
26. Ballaré CL (2014) Light regulation of plant defense. Annu Rev Plant Biol 65(1):335–363. https://doi.org/10.1146/annurev-arplant-050213-040145
27. Siciliano V, Genre A, Balestrini R, Cappellazzo G, deWit PJGM, Bonfante P (2007) Transcriptome analysis of arbuscular mycorrhizal roots during development of the prepenetration apparatus. Plant Physiol 144 (3):1455–1466. https://doi.org/10.1104/pp.107.097980
28. Blilou I, Ocampo JA, García-Garrido JM (1999) Resistance of pea roots to endomycorrhizal fungus or rhizobium correlates with enhanced levels of endogenous salicylic acid. J Exp Bot 50(340):1663–1668. https://doi.org/10.1093/jxb/50.340.1663
29. Blilou I, Ocampo JA, García-Garrido JM, Garcia-Garrido JM (2000) Induction of Ltp (lipid transfer protein) and pal (phenylalanine ammonia-lyase) gene expression in rice roots colonized by the arbuscular mycorrhizal fungus Glomus mosseae. J Exp Bot 51 (353):1969–1977. https://doi.org/10.1093/jexbot/51.353.1969
30. Gutjahr C, Paszkowski U (2009) Weights in the balance: jasmonic acid and salicylic acid signaling in root-biotroph interactions. Mol Plant-Microbe Interact 22(7):763–772. https://doi.org/10.1094/MPMI-22-7-0763
31. Hause B, Maier W, Miersch O, Kramell R, Strach D (2002) Induction of Jasmonate biosynthesis in Arbuscular Mycorrhizal barley roots. Plant Physiol 130(3):1213–1220. https://doi.org/10.1104/pp.006007
32. Stumpe M, Carsjens J-G, Stenzel I, Göbel C, Lang I, Pawlowski K et al (2005) Lipid metabolism in arbuscular mycorrhizal roots of Medicago truncatula. Phytochemistry 66 (7):781–791. https://doi.org/10.1016/j.phytochem.2005.01.020
33. Meixner C, Ludwig-Müller J, Miersch O, Gresshoff P, Staehelin C, Vierheilig H (2005) Lack of mycorrhizal autoregulation and phytohormonal changes in the supernodulating soybean mutant nts1007. Planta 222 (4):709–715. https://doi.org/10.1007/s00425-005-0003-4
34. Isayenkov S, Mrosk C, Stenzel I, Strack D, Hause B (2005) Suppression of allene oxide cyclase in hairy roots of Medicago truncatula reduces Jasmonate levels and the degree of Mycorrhization with Glomus intraradices. Plant Physiol 139(3):1401–1410. https://doi.org/10.1104/pp.105.069054
35. León-Morcillo RJ, Ángel J, Martín-Rodríguez, Vierheilig H, Ocampo JA, García-

Garrido JM (2012) Late activation of the 9-oxylipin pathway during arbuscular mycorrhiza formation in tomato and its regulation by jasmonate signalling. J Exp Bot 63 (10):3545–3558. https://doi.org/10.1093/jxb/ers010

36. Tejeda-Sartorius M, Martínez De La Vega O, Délano-Frier JP (2008) Jasmonic acid influences mycorrhizal colonization in tomato plants by modifying the expression of genes involved in carbohydrate partitioning. Physiol Plant 133(2):339–353. https://doi.org/10.1111/j.1399-3054.2008.01081.x
37. Hause B, Mrosk C, Isayenkov S, Strack D (2007) Jasmonates in arbuscular mycorrhizal interactions. Phytochemistry 68(1):101–110. https://doi.org/10.1016/j.phytochem.2006.09.025
38. Herrera-Medina MJ, Tamayo MI, Vierheilig H, Ocampo JA, García-Garrido JM (2008) The jasmonic acid signalling pathway restricts the development of the arbuscular mycorrhizal association in tomato. J Plant Growth Regul 27(3):221–230. https://doi.org/10.1007/s00344-008-9049-4
39. Regvar M, Gogala N, Zalar P (1996) Effect of jasmonic acid on mycorrhizal Allium sativum. New Phytol 134(134):703–707. https://doi.org/10.1111/j.1469-8137.1996.tb04936.x
40. Kiers ET, Adler LS, Grman EL, van der Heijden MGA (2010) Manipulating the jasmonate response: how do methyl jasmonate additions mediate characteristics of aboveground and belowground mutualisms? Funct Ecol 24(2):434–443. https://doi.org/10.1111/j.1365-2435.2009.01625.x
41. Landgraf R, Schaarschmidt S, Hause B (2012) Repeated leaf wounding alters the colonization of Medicago truncatula roots by beneficial and pathogenic microorganisms. Plant Cell Environ 35(7):1344–1357. https://doi.org/10.1111/j.1365-3040.2012.02495.x
42. Gutjahr C, Siegler H, Haga K, Iino M, Paszkowski U (2015) Full establishment of arbuscular mycorrhizal symbiosis in rice occurs independently of enzymatic jasmonate biosynthesis. PLoS One 10(4):e0123422. https://doi.org/10.1371/journal.pone.0123422
43. Ludwig-Müller J, Bennett RN, García-Garrido JM, Piché Y, Vierheilig H (2002) Reduced arbuscular mycorrhizal root colonization in Tropaeolum majus and Carica papaya after jasmonic acid application can not be attributed to increased glucosinolate levels. J Plant Physiol 159(5):517–523. https://doi.org/10.1078/0176-1617-00731
44. Riedel T, Groten K, Baldwin IT (2008) Symbiosis between Nicotiana attenuata and Glomus intraradices: ethylene plays a role, jasmonic acid does not. Plant Cell Environ 31(9):1203–1213. https://doi.org/10.1111/j.1365-3040.2008.01827.x
45. Fernández I, Merlos M, López-Ráez JA, Martínez-Medina A, Ferrol N, Azcón C et al (2014) Defense related Phytohormones regulation in Arbuscular Mycorrhizal symbioses depends on the partner genotypes. J Chem Ecol 40(7):791–803. https://doi.org/10.1007/s10886-014-0473-6
46. Yang Z-B, He C, Ma Y, Herde M, Ding Z (2017) Jasmonic acid enhances Al-induced root growth inhibition. Plant Physiol 173 (2):1420–1433. https://doi.org/10.1104/pp.16.01756
47. Genre A, Bonfante P (1998) Actin versus tubulin configuration in arbuscule-containing cells from mycorrhizal tobacco roots. New Phytol 140(4):745–752. https://doi.org/10.1046/j.1469-8137.1998.00314.x
48. Blancaflor EB, Zhao L, Harrison MJ (2001) Microtubule organization in root cells of Medicago truncatula during development of an arbuscular mycorrhizal symbiosis with Glomus versiforme. Protoplasma 217 (4):154–165. https://doi.org/10.1007/BF01283396
49. Manthey K, Krajinski F, Hohnjec N, Firnhaber C, Pühler A, Perlick AM, Küster H (2004) Transcriptome profiling in root nodules and Arbuscular Mycorrhiza identifies a collection of novel genes induced during Medicago truncatula root Endosymbioses. Mol Plant-Microbe Interact 17 (10):1063–1077. https://doi.org/10.1094/MPMI.2004.17.10.1063
50. Koda Y (1997) Possible involvement of jasmonates in various morphogenic events. Physiol Plant 100(3):639–646. https://doi.org/10.1111/j.1399-3054.1997.tb03070.x
51. Babst BA, Ferrieri RA, Gray DW, Lerdau M, Schlyer DJ, Schueller M et al (2005) Jasmonic acid induces rapid changes in carbon transport and partitioning in Populus. New Phytol 167:63–72. https://doi.org/10.1111/j.1469-8137.2005.01388.x
52. Gómez S, Ferrieri RA, Schueller M, Orians CM (2010) Methyl jasmonate elicits rapid changes in carbon and nitrogen dynamics in tomato. New Phytol 188(3):835–844. https://doi.org/10.1111/j.1469-8137.2010.03414.x
53. Hause B, Stenzel I, Miersch O, Maucher H, Kramell R, Ziegler J, Wasternack C (2000) Tissue-specific oxylipin signature of tomato

flowers: allene oxide cyclase is highly expressed in distinct flower organs and vascular bundles. Plant J 24(1):113–126. https://doi.org/10.1046/j.1365-313X.2000.00861.x

54. Maucher H, Hause B, Feussner I, Ziegler J, Wasternack C (2001) Allene oxide synthases of barley (Hordeum vulgare cv. Salome): tissue specific regulation in seedling development. Plant J 21(2):199–213. https://doi.org/10.1046/j.1365-313x.2000.00669.x
55. Sánchez-Bel P, Sanmartín N, Pastor V, Mateu D, Cerezo M, Vidal-Albalat A et al (2018) Mycorrhizal tomato plants fine tunes the growth-defence balance upon N depleted root environments. Plant Cell Environ 41 (2):406–420. https://doi.org/10.1111/pce.13105
56. Schaarschmidt S, Roitsch T, Hause B (2006) Arbuscular mycorrhiza induces gene expression of the apoplastic invertase LIN6 in tomato (Lycopersicon esculentum) roots. J Exp Bot 57(15):4015–4023. https://doi.org/10.1093/jxb/erl172
57. Thoma I, Loeffler C, Sinha AK, Gupta M, Krischke M, Steffan B et al (2003) Cyclopentenone isoprostanes induced by reactive oxygen species trigger defense gene activation and phytoalexin accumulation in plants. Plant J 34(3):363–375. https://doi.org/10.1046/j.1365-313X.2003.01730.x
58. Daher Z, Recorbet G, Solymosi K, Wienkoop S, Mounier A, Morandi D et al (2017) Changes in plastid proteome and structure in arbuscular mycorrhizal roots display a nutrient starvation signature. Physiol Plant 159(1):13–29. https://doi.org/10.1111/ppl.12505
59. Harrison MJ, Dixon RA (1994) Spatial patterns of expression of flavonoid/isoflavonoid pathway genes during interactions between roots of Medicago truncatula and the mycorrhizal fungus Glomus versiforme. Plant J 6 (1):9–20. https://doi.org/10.1046/j.1365-313X.1994.6010009.x
60. Vellosillo T, Martinez M, Lopez MA, Vicente J, Cascon T, Dolan L et al (2007) Oxylipins produced by the 9-Lipoxygenase pathway in Arabidopsis regulate lateral root development and defense responses through a specific signaling Cascade. Plant Cell 19 (3):831–846. https://doi.org/10.1105/tpc.106.046052
61. López-Ráez JA, Verhage A, Fernández I, García JM, Azcón-Aguilar C, Flors V, Pozo MJ (2010) Hormonal and transcriptional profiles highlight common and differential host responses to arbuscular mycorrhizal fungi and the regulation of the oxylipin pathway. J Exp Bot 61(10):2589–2601. https://doi.org/10.1093/jxb/erq089
62. Brewin NJ (1991) Development of the legume root nodule. Annu Rev Cell Biol 7 (1):191–226. https://doi.org/10.1146/annurev.cb.07.110191.001203
63. Suzuki A, Suriyagoda L, Shigeyama T, Tominaga A, Sasaki M, Hiratsuka Y et al (2011) Lotus japonicus nodulation is photomorphogenetically controlled by sensing the red/far red (R/FR) ratio through jasmonic acid (JA) signaling. Proc Natl Acad Sci 108 (40):16837–16842. https://doi.org/10.1073/pnas.1105892108
64. Nakagawa T, Kawaguchi M (2006) Shoot-applied MeJA suppresses root nodulation in Lotus japonicus. Plant Cell Physiol 47 (1):176–180. https://doi.org/10.1093/pcp/pci222
65. Sun J, Cardoza V, Mitchell DM, Bright L, Oldroyd G, Harris JM (2006) Crosstalk between jasmonic acid, ethylene and nod factor signaling allows integration of diverse inputs for regulation of nodulation. Plant J 46(6):961–970. https://doi.org/10.1111/j.1365-313X.2006.02751.x
66. Seo HS, Li J, Lee S-Y, Yu J-W, Kim K-H, Lee S-H, Paek I-J, Paek NC (2007) The Hypernodulating nts mutation induces Jasmonate synthetic pathway in soybean leaves. Mol Cells 24(2):185–193
67. Oldroyd GED, Downie JA (2008) Coordinating nodule morphogenesis with Rhizobial infection in legumes. Annu Rev Plant Biol 59 (1):519–546. https://doi.org/10.1146/annurev.arplant.59.032607.092839
68. Rosas S, Soria R, Correa N, Abdala G (1998) Jasmonic acid stimulates the expression of nod genes in rhizobium. Plant Mol Biol 38:1161–1168. https://doi.org/10.1023/A:1006064807870
69. Mabood F, Souleimanov A, Khan W, Smith DL (2006) Jasmonates induce nod factor production by Bradyrhizobium japonicum. Plant Physiol Biochem 44(11):759–765. https://doi.org/10.1016/j.plaphy.2006.10.025
70. Cooper JE (2004) Multiple responses of rhizobia to flavonoids during legume root infection. Adv Bot Res 41:1–62. https://doi.org/10.1016/S0065-2296(04)41001-5
71. Zhang N, Venkateshwaran M, Boersma M, Harms A, Howes-Podoll M, Den Os D et al (2012) Metabolomic profiling reveals suppression of oxylipin biosynthesis during the early stages of legume-rhizobia symbiosis.

FEBS Lett 586(19):3150–3158. https://doi.org/10.1016/j.febslet.2012.06.046

72. Kouchi H, Shimomura K, Hata S, Hirota A, Wu GJ, Kumagai H et al (2004) Large-scale analysis of gene expression profiles during early stages of root nodule formation in a model legume, Lotus japonicus. DNA Res 11(4):263–274. https://doi.org/10.1093/dnares/11.4.263
73. Perlick AM, Albus U, Stavridis T, Frühling M, Küster H, Pühler A (1996) The Vicia faba Lipoxygenase gene VfLOX1 is expressed in the root nodule parenchyma. Mol Plant-Microbe Interact 9(9):860–863. https://doi.org/10.1094/MPMI-9-0860
74. Bueno P, Soto MJ, Rodríguez-Rosales MP, Sanjuan J, Olivares J, Donaire JP (2001) Time-course of lipoxygenase, antioxidant enzyme activities and H2O2 accumulation during the early stages of rhizobium–legume symbiosis. New Phytol 152(1):91–96. https://doi.org/10.1046/j.0028-646x.2001.00246.x
75. Mohammadi M, Karr AL (2003) Induced lipoxygenases in soybean root nodules. Plant Sci 164(4):471–479. https://doi.org/10.1016/S0168-9452(02)00431-4
76. Costanzo ME, Andrade A, Del Carmen Tordable M, Cassán F, Abdala G (2012) Production and function of jasmonates in nodulated roots of soybean plants inoculated with Bradyrhizobium japonicum. Arch Microbiol 194(10):837–845. https://doi.org/10.1007/s00203-012-0817-y
77. Demchenko K, Zdyb A, Feussner I, Pawlowski K (2012) Analysis of the subcellular localisation of lipoxygenase in legume and actinorhizal nodules. Plant Biol 14(1):56–63. https://doi.org/10.1111/j.1438-8677.2011.00480.x
78. Zdyb A, Salgado MG, Demchenko KN, Brenner WG, Płaszczyca M, Stumpe M et al (2018) Allene oxide synthase, allene oxide cyclase and jasmonic acid levels in Lotus japonicus nodules. PLoS One 13(1):1–16. https://doi.org/10.1371/journal.pone.0190884
79. Zdyb A, Demchenko K, Heumann J, Mrosk C, Grzeganek P, Göbel C et al (2011) Jasmonate biosynthesis in legume and actinorhizal nodules. New Phytol 189(2):568–579. https://doi.org/10.1111/j.1469-8137.2010.03504.x
80. Pauwels L, Goossens A (2011) The JAZ proteins: a crucial Interface in the Jasmonate signaling Cascade. Plant Cell 23(9):3089–3100. https://doi.org/10.1105/tpc.111.089300
81. Li Y, Xu M, Wang N, Li Y (2015) A JAZ protein in Astragalus sinicus interacts with a leghemoglobin through the TIFY domain and is involved in nodule development and nitrogen fixation. PLoS One 10(10):e0139964. https://doi.org/10.1371/journal.pone.0139964
82. Wittenberg JB, Bergersen FJ, Appleby CA, Turner GL (1974) The role of leghemoglobin in nitrogen fixation by bacteroids isolated from soybean root nodules. J Biol Chem 29(13):4057–4066. http://www.jbc.org/content/249/13/4057.abstract
83. Ott T, van Dongen JT, Günther C, Krusell L, Desbrosses G, Vigeolas H et al (2005) Symbiotic Leghemoglobins are crucial for nitrogen fixation in legume root nodules but not for general plant growth and development. Curr Biol 15(6):531–535. https://doi.org/10.1016/j.cub.2005.01.042
84. Reid DE, Ferguson BJ, Hayashi S, Lin Y-H, Gresshoff PM (2011) Molecular mechanisms controlling legume autoregulation of nodulation. Ann Bot 108(5):789–795. https://doi.org/10.1093/aob/mcr205
85. Krusell L, Madsen LH, Sato S, Aubert G, Genua A, Szczyglowski K et al (2002) Shoot control of root development and nodulation is mediated by a receptor-like kinase. Nature 420:422–426. https://doi.org/10.1038/nature01207
86. Nishimura R, Hayashi M, Wu G-J, Kouchi H, Imaizumi-Anraku H, Murakami Y et al (2002) HAR1 mediates systemic regulation of symbiotic organ development. Nature 420:426. https://doi.org/10.1038/nature01231
87. Searle IR, Men AE, Laniya TS, Buzas DM, Iturbe-Ormaetxe I, Carroll BJ, Gresshoff PM (2003) Long-distance signaling in nodulation directed by a CLAVATA1-like receptor kinase. Science 299(5603):109–112. https://doi.org/10.1126/science.1077937
88. Schnabel E, Journet EP, De Carvalho-Niebel-F, Duc G, Frugoli J (2005) The Medicago truncatula SUNN gene encodes a CLV1-like leucine-rich repeat receptor kinase that regulates nodule number and root length. Plant Mol Biol 58(6):809–822. https://doi.org/10.1007/s11103-005-8102-y
89. Okamoto S, Ohnishi E, Sato S, Takahashi H, Nakazono M, Tabata S, Kawaguchi M (2009) Nod factor/nitrate-induced CLE genes that drive HAR1-mediated systemic regulation of nodulation. Plant Cell Physiol 50(1):67–77. https://doi.org/10.1093/pcp/pcn194
90. Mortier V, Den Herder G, Whitford R, Van de Velde W, Rombauts S, D'haeseleer K et al

(2010) CLE peptides control Medicago truncatula nodulation locally and systemically. Plant Physiol 153(1):222–237. https://doi.org/10.1104/pp.110.153718
91. Kinkema M, Gresshoff PM (2008) Investigation of downstream signals of the soybean autoregulation of nodulation receptor kinase GmNARK. Mol Plant-Microbe Interact 21 (10):1337–1348. https://doi.org/10.1094/MPMI-21-10-1337
92. Tsikou D, Yan Z, Holt DB, Abel NB, Reid DE, Madsen LH et al (2018) Systemic control of legume susceptibility to rhizobial infection by a mobile microRNA. Science 6907(8): eaat6907. https://doi.org/10.1126/science.aat6907
93. Magori S, Kawaguchi M (2010) Analysis of two potential long-distance signaling molecules, LjCLE-RS1/2 and jasmonic acid, in a hypernodulating mutant too much love. Plant Signal Behav 5(4):403–405. http://www.ncbi.nlm.nih.gov/pmc/articles/PMC2958591/
94. Dombrecht B, Xue GP, Sprague SJ, Kirkegaard JA, Ross JJ, Reid JB et al (2007) MYC2 differentially modulates diverse jasmonate-dependent functions in Arabidopsis. Plant Cell 19(7):2225–2245. https://doi.org/10.1105/tpc.106.048017
95. Caba JM, Centeno ML, Fernández B, Gresshoff PM, Ligero F (2000) Inoculation and nitrate alter phytohormone levels in soybean roots: differences between a supernodulating mutant and the wild type. Planta 211 (1):98–104. https://doi.org/10.1007/s004250000265
96. Van Noorden GE, Ross JJ, Reid JB, Rolfe BG, Mathesius U (2006) Defective long-distance Auxin transport regulation in the Medicago truncatula super numeric nodules mutant. Plant Physiol 140(4):1494 LP–1491506. http://www.plantphysiol.org/content/140/4/1494.abstract
97. Heath KD, Lau JA (2011) Herbivores alter the fitness benefits of a plant–rhizobium mutualism. Acta Oecol 37(2):87–92. https://doi.org/10.1016/j.actao.2010.12.002
98. Perez-Moreno J, Read DJ (2000) Mobilization and transfer of nutrients from litter to tree seedlings via the vegetative mycelium of ectomycorrhizal plants. New Phytol 145 (2):301–309. https://doi.org/10.1046/j.1469-8137.2000.00569.x
99. Tibbett M, Sanders FE (2002) Ectomycorrhizal symbiosis can enhance plant nutrition through improved access to discrete organic nutrient patches of high resource quality. Ann Bot 89(6):783–789. http://www.jstor.org/stable/42771516
100. Luo Z-B, Janz D, Jiang X, Gobel C, Wildhagen H, Tan Y et al (2009) Upgrading root physiology for stress tolerance by ectomycorrhizas: insights from metabolite and transcriptional profiling into reprogramming for stress anticipation. Plant Physiol 151 (4):1902–1917. https://doi.org/10.1104/pp.109.143735
101. Clemmensen KE, Bahr A, Ovaskainen O, Dahlberg A, Ekblad A, Wallander H et al (2013) Roots and associated fungi drive long-term carbon sequestration in boreal Forest. Science 339(6127):1615–1618. https://doi.org/10.1126/science.1231923
102. Lindahl BD, Tunlid A (2014) Ectomycorrhizal fungi – potential organic matter decomposers, yet not saprotrophs. New Phytol 205 (4):1443–1447. https://doi.org/10.1111/nph.13201
103. Peterson RL, Massicotte HB (2004) Exploring structural definitions of mycorrhizas, with emphasis on nutrient-exchange interfaces. Can J Bot 82(8):1074–1088. https://doi.org/10.1139/b04-071
104. Kohler A, Kuo A, Nagy LG, Morin E, Barry KW, Buscot F et al (2015) Convergent losses of decay mechanisms and rapid turnover of symbiosis genes in mycorrhizal mutualists. Nat Genet 47:410. https://doi.org/10.1038/ng.3223
105. Plett JM, Kemppainen M, Kale SD, Kohler A, Legué V, Brun A et al (2011) A secreted effector protein of Laccaria bicolor is required for symbiosis development. Curr Biol 21 (14):1197–1203. https://doi.org/10.1016/j.cub.2011.05.033
106. Plett JM, Daguerre Y, Wittulsky S, Vayssieres A, Deveau A, Melton SJ et al (2014) Effector MiSSP7 of the mutualistic fungus Laccaria bicolor stabilizes the Populus JAZ6 protein and represses jasmonic acid (JA) responsive genes. Proc Natl Acad Sci 111(22):8299–8304. https://doi.org/10.1073/pnas.1322671111
107. Plett JM, Khachane A, Ouassou M, Sundberg B, Kohler A, Martin F (2014b) Ethylene and jasmonic acid act as negative modulators during mutualistic symbiosis between Laccaria bicolor and Populus roots. New Phytol 202(1):270–286. https://doi.org/10.1111/nph.12655
108. Sasse J, Martinoia E, Northen T (2018) Feed your friends: do plant exudates shape the root microbiome? Trends Plant Sci 23(1):25–41. https://doi.org/10.1016/j.tplants.2017.09.003

109. Vandenkoornhuyse P, Quaiser A, Duhamel M, Le Van A, Dufresne A (2015) The importance of the microbiome of the plant holobiont. New Phytol 206 (4):1196–1206. https://doi.org/10.1111/nph.13312
110. Goh C-H, Veliz Vallejos DF, Nicotra AB, Mathesius U (2013) The impact of beneficial plant-associated microbes on plant phenotypic plasticity. J Chem Ecol 39(7):826–839. https://doi.org/10.1007/s10886-013-0326-8
111. Pieterse CMJ, Zamioudis C, Berendsen RL, Weller DM, Van Wees SCM, Bakker PAHM (2014) Induced systemic resistance by beneficial microbes. Annu Rev Phytopathol 52 (1):347–375. https://doi.org/10.1146/annurev-phyto-082712-102340
112. Shoresh M, Harman GE, Mastouri F (2010) Induced systemic resistance and plant responses to fungal biocontrol agents. Annu Rev Phytopathol 48(1):21–43. https://doi.org/10.1146/annurev-phyto-073009-114450
113. Walters DR, Ratsep J, Havis ND (2013) Controlling crop diseases using induced resistance: challenges for the future. J Exp Bot 64 (5):1263–1280. https://doi.org/10.1093/jxb/ert026
114. Haney CH, Samuel BS, Bush J, Ausubel FM, Hospital MG (2015) Associations with rhizosphere bacteria can confer an adaptive advantage to plants. Nat Plants 1(6):1–22. https://doi.org/10.1038/nplants.2015.51
115. Wintermans PCA, Bakker PAHM, Pieterse CMJ (2016) Natural genetic variation in Arabidopsis for responsiveness to plant growth-promoting rhizobacteria. Plant Mol Biol 90 (6):623–634. https://doi.org/10.1007/s11103-016-0442-2
116. Iavicoli A, Boutet E, Buchala A, Métraux J-P (2003) Induced systemic resistance in Arabidopsis thaliana in response to root inoculation with Pseudomonas fluorescens CHA0. Mol Plant-Microbe Interact 16(10):851–858. https://doi.org/10.1094/MPMI.2003.16.10.851
117. Pineda A, Soler R, Weldegergis BT, Shimwela MM, Van Loon JJA, Dicke M (2013) Non-pathogenic rhizobacteria interfere with the attraction of parasitoids to aphid-induced plant volatiles via jasmonic acid signalling. Plant Cell Environ 36(2):393–404. https://doi.org/10.1111/j.1365-3040.2012.02581.x
118. Weller DM, Raaijmakers JM, Gardener BBM, Thomashow LS (2002) Microbial populations responsible for specific soil suppressiveness to plant pathogens. Annu Rev Phytopathol 40(1):309–348. https://doi.org/10.1146/annurev.phyto.40.030402.110010
119. Raaijmakers JM, Paulitz TC, Steinberg C, Alabouvette C, Moënne-Loccoz Y (2009) The rhizosphere: a playground and battlefield for soilborne pathogens and beneficial microorganisms. Plant Soil 321(1):341–361. https://doi.org/10.1007/s11104-008-9568-6
120. Mendes R, Kruijt M, de Bruijn I, Dekkers E, van der Voort M, Schneider JHM, Piceno YM, DeSantis TZ, Andersen GL, Bakker P, Raaijmakers J (2012) Deciphering the Rhizosphere microbiome. Science 1097 (2011):1097–1100. https://doi.org/10.1126/science.1203980
121. Carvalhais LC, Dennis PG, Badri DV, Tyson GW, Vivanco JM, Schenk PM (2013) Activation of the Jasmonic acid plant defence pathway alters the composition of Rhizosphere bacterial communities. PLoS One 8(2): e56457. https://doi.org/10.1371/journal.pone.0056457
122. Liu H, Carvalhais LC, Schenk PM, Dennis PG (2017) Effects of jasmonic acid signalling on the wheat microbiome differ between body sites. Sci Rep 7:41766. https://doi.org/10.1038/srep41766
123. Carvalhais LC, Dennis PG, Badri DV, Kidd BN, Vivanco JM, Schenk PM (2015) Linking Jasmonic acid signaling, root exudates, and Rhizosphere microbiomes. Mol Plant-Microbe Interact 28(9):1049–1058. https://doi.org/10.1094/MPMI-01-15-0016-R
124. Miché L, Battistoni F, Gemmer S, Belghazi M, Reinhold-Hurek B (2006) Upregulation of jasmonate-inducible defense proteins and differential colonization of roots of Oryza sativa cultivars with the endophyte Azoarcus sp. Mol Plant-Microbe Interact 19 (5):502–511. https://doi.org/10.1094/MPMI-19-0502
125. Martínez-Medina A, Appels FVW, van Wees SCM (2017) Impact of salicylic acid- and jasmonic acid-regulated defences on root colonization by Trichoderma harzianum T-78. Plant Signal Behav 12(8):e1345404. https://doi.org/10.1080/15592324.2017.1345404
126. Varma A, Verma S, Sudha, Sahay N, Bütehorn B, Franken P (1999) Piriformospora indica, a cultivable plant-growth-promoting root endophyte. Appl Environ Microbiol 65(6):2741–2744
127. Waller F, Achatz B, Baltruschat H, Fodor J, Becker K, Fischer M et al (2005) The

endophytic fungus Piriformospora indica reprograms barley to salt-stress tolerance, disease resistance, and higher yield. Proc Natl Acad Sci 102(38):13386–13391. https://doi.org/10.1073/pnas.0504423102

128. Jacobs S, Zechmann B, Molitor A, Trujillo M, Petutschnig E, Lipka V et al (2011) Broad-Spectrum suppression of innate immunity is required for colonization of Arabidopsis roots by the fungus Piriformospora indica. Plant Physiol 156(2):726–740. https://doi.org/10.1104/pp.111.176446
129. Zuccaro A, Lahrmann U, Güldener U, Langen G, Pfiffi S, Biedenkopf D et al (2011) Endophytic life strategies decoded by genome and transcriptome analyses of the mutualistic root symbiont Piriformospora indica. PLoS Pathog 7(10):e1002290. https://doi.org/10.1371/journal.ppat.1002290
130. Baker K, Cook RJ (1974) Biological control of plant pathogens. W.H. Freeman and Company, San Francisco
131. Conrath U, Beckers GJM, Flors V, García-Agustín P, Jakab G, Mauch F et al (2006) Priming: getting ready for Battle. Mol Plant-Microbe Interact 19(10):1062–1071. https://doi.org/10.1094/MPMI-19-1062
132. Van Wees SCM, Luijendijk M, Smoorenburg I, van Loon LC, Pieterse CMJ (1999) Rhizobacteria-mediated induced systemic resistance (ISR) in Arabidopsis is not associated with a direct effect on expression of known defense-related genes but stimulates the expression of the jasmonate-inducible gene Atvsp upon challenge. Plant Mol Biol 41(4):537–549. https://doi.org/10.1023/A:1006319216982
133. Pieterse CMJ, Van Wees SCM, Ton J, Van Pelt JA, Van Loon LC (2002) Signalling in rhizobacteria-induced systemic resistance in Arabidopsis thaliana. Plant Biol 4 (5):535–544. https://doi.org/10.1055/s-2002-35441
134. Pieterse CMJ, Van Pelt JA, Ton J, Parchmann S, Mueller MJ, Buchala AJ et al (2000) Rhizobacteria-mediated induced systemic resistance (ISR) in Arabidopsis requires sensitivity to jasmonate and ethylene but is not accompanied by an increase in their production. Physiol Mol Plant Pathol 57 (3):123–134. https://doi.org/10.1006/pmpp.2000.0291
135. Djonović S, Vargas WA, Kolomiets MV, Horndeski M, Wiest A, Kenerley CM (2007) A Proteinaceous elicitor Sm1 from the beneficial fungus Trichoderma virens is required for induced systemic resistance in maize. Plant Physiol 145(3):875–889. https://doi.org/10.1104/pp.107.103689
136. Martínez-Medina A, Roldán A, Albacete A, Pascual JA (2011) The interaction with arbuscular mycorrhizal fungi or Trichoderma harzianum alters the shoot hormonal profile in melon plants. Phytochemistry 72 (2–3):223–229. https://doi.org/10.1016/j.phytochem.2010.11.008
137. Shavit R, Ofek-Lalzar M, Burdman S, Morin S (2013) Inoculation of tomato plants with rhizobacteria enhances the performance of the phloem-feeding insect Bemisia tabaci. Front Plant Sci 4(8):306. https://doi.org/10.3389/fpls.2013.00306
138. Dean JM, Mescher MC, De Moraes CM (2014) Plant dependence on rhizobia for nitrogen influences induced plant defenses and herbivore performance. Int J Mol Sci 15 (1):1466–1480. https://doi.org/10.3390/ijms15011466
139. Van Wees SC, Van der Ent S, Pieterse CM (2008) Plant immune responses triggered by beneficial microbes. Curr Opin Plant Biol 11 (4):443–448. https://doi.org/10.1016/j.pbi.2008.05.005
140. Pieterse CMJ, Van Wees SCM, Van Pelt JA, Knoester M, Laan R, Gerrits H et al (1998) A novel signaling pathway controlling induced systemic resistance in Arabidopsis. Plant Cell 10(9):1571–1580. https://doi.org/10.1105/tpc.10.9.1571
141. Knoester M, Pieterse CMJ, Bol JF, Van Loon LC (1999) Systemic resistance in Arabidopsis induced by Rhizobacteria requires ethylene-dependent signaling at the site of application. Mol Plant-Microbe Interact 12(8):720–727. https://doi.org/10.1094/MPMI.1999.12.8.720
142. Yan Z, Reddy MS, Ryu C-M, McInroy JA, Wilson M, Kloepper JW (2002) Induced systemic protection against tomato late blight elicited by plant growth-promoting Rhizobacteria. Phytopathology 92 (12):1329–1333. https://doi.org/10.1094/PHYTO.2002.92.12.1329
143. Ryu C-M, Murphy JF, Mysore KS, Kloepper JW (2004) Plant growth-promoting rhizobacteria systemically protect Arabidopsis thaliana against cucumber mosaic virus by a salicylic acid and NPR1-independent and jasmonic acid-dependent signaling pathway. Plant J 39(3):381–392. https://doi.org/10.1111/j.1365-313X.2004.02142.x
144. Shoresh M, Yedidia I, Chet I (2005) Involvement of Jasmonic acid/ethylene signaling pathway in the systemic resistance induced in cucumber by Trichoderma asperellum T203.

Phytopathology 95(1):76–84. https://doi.org/10.1094/PHYTO-95-0076
145. Ahn I-P, Lee S-W, Suh S-C (2007) Rhizobacteria-induced priming in Arabidopsis is dependent on ethylene, Jasmonic acid, and NPR1. Mol Plant-Microbe Interact 20(7):759–768. https://doi.org/10.1094/MPMI-20-7-0759
146. De Vleesschauwer D, Bakker PAHM, Djavaheri M, Hofte M (2008) Pseudomonas fluorescens WCS374r-induced systemic resistance in rice against Magnaporthe oryzae is based on pseudobactin-mediated priming for a salicylic acid-repressible multifaceted defense response. Plant Physiol 148:1996–2012. https://doi.org/10.1104/pp.108.127878
147. Hase S, Takahashi S, Takenaka S, Nakaho K, Arie T, Seo S et al (2008) Involvement of jasmonic acid signalling in bacterial wilt disease resistance induced by biocontrol agent Pythium oligandrum in tomato. Plant Pathol 57(5):870–876. https://doi.org/10.1111/j.1365-3059.2008.01858.x
148. Hossain MM, Sultana F, Kubota M, Hyakumachi M (2008) Differential inducible defense mechanisms against bacterial speck pathogen in Arabidopsis thaliana by plant-growth-promoting-fungus Penicillium sp. GP16-2 and its cell free filtrate. Plant Soil 304(1):227–239. https://doi.org/10.1007/s11104-008-9542-3
149. Korolev N, Rav David D, Elad Y (2008) The role of phytohormones in basal resistance and Trichoderma-induced systemic resistance to Botrytis cinerea in Arabidopsis thaliana. Biol Control 53(4):667–683. https://doi.org/10.1007/s10526-007-9103-3
150. Stein E, Molitor A, Kogel K-H, Waller F (2008) Systemic resistance in Arabidopsis conferred by the Mycorrhizal fungus Piriformospora indica requires Jasmonic acid signaling and the cytoplasmic function of NPR1. Plant Cell Physiol 49(11):1747–1751. https://doi.org/10.1093/pcp/pcn147
151. Weller DM, Mavrodi DV, Van Pelt JA, Pieterse CMJ, Van Loon LC, Bakker PA (2012) Induced systemic resistance in Arabidopsis thaliana against pseudomonas syringae pv. Tomato by 2,4-diacetylphloroglucinol-producing Pseudomonas fluorescens. Phytopathology 102(4):403–412. https://doi.org/10.1094/PHYTO-08-11-0222
152. Pangesti N, Reichelt M, van de Mortel JE, Kapsomenou E, Gershenzon J, van Loon JJA et al (2016) Jasmonic acid and ethylene signaling pathways regulate Glucosinolate levels in plants during Rhizobacteria-induced systemic resistance against a leaf-chewing herbivore. J Chem Ecol 42(12):1212–1225. https://doi.org/10.1007/s10886-016-0787-7
153. Van der Ent S, Verhagen BWM, Van Doorn R, Bakker D, Verlaan MG, Pel MJC et al (2008) MYB72 is required in early signaling steps of Rhizobacteria-induced systemic resistance in Arabidopsis. Plant Physiol 146(3):1293–1304. https://doi.org/10.1104/pp.107.113829
154. Segarra G, Van Der Ent S, Trillas I, Pieterse CMJ (2009) MYB72, a node of convergence in induced systemic resistance triggered by a fungal and a bacterial beneficial microbe. Plant Biol 11(1):90–96. https://doi.org/10.1111/j.1438-8677.2008.00162.x
155. Audenaert K, Pattery T, Cornelis P, Höfte M (2002) Induction of systemic resistance to Botrytis cinerea in tomato by Pseudomonas aeruginosa 7NSK2: role of salicylic acid, pyochelin, and pyocyanin. Mol Plant-Microbe Interact 15(11):1147–1156. https://doi.org/10.1094/MPMI.2002.15.11.1147
156. Tjamos SE, Flemetakis E, Paplomatas EJ, Katinakis P (2005) Induction of resistance to Verticillium dahliae in Arabidopsis thaliana by the biocontrol agent K-165 and pathogenesis-related proteins gene expression. Mol Plant-Microbe Interact 18(6):555–561. https://doi.org/10.1094/MPMI-18-0555
157. Van de Mortel JE, de Vos RCH, Dekkers E, Pineda A, Guillod L, Bouwmeester K et al (2012) Metabolic and transcriptomic changes induced in Arabidopsis by the Rhizobacterium Pseudomonas fluorescens SS101. Plant Physiol 160(4):2173–2188. https://doi.org/10.1104/pp.112.207324
158. Conn VM, Walker AR, Franco CMM (2008) Endophytic Actinobacteria induce defense pathways in Arabidopsis thaliana. Mol Plant-Microbe Interact 21(2):208–218. https://doi.org/10.1094/MPMI-21-2-0208
159. Contreras-Cornejo HA, Macías-Rodríguez L, Beltrán-Peña E, Herrera-Estrella A, López-Bucio J (2011) Trichoderma-induced plant immunity likely involves both hormonal- and camalexin-dependent mechanisms in Arabidopsis thaliana and confers resistance against necrotrophic fungus Botrytis cinerea. Plant Signal Behav 6(10):1554–1563. https://doi.org/10.4161/psb.6.10.17443
160. Mathys J, De Cremer K, Timmermans P, Van Kerkhove S, Lievens B, Vanhaecke M et al (2012) Genome-wide characterization of ISR induced in Arabidopsis thaliana by Trichoderma hamatum T382 against Botrytis

cinerea infection. Front Plant Sci 3:108. https://doi.org/10.3389/fpls.2012.00108

161. Martínez-Medina A, Fernández I, Sánchez-Guzmán MJ, Jung SC, Pascual JA, Pozo MJ (2013) Deciphering the hormonal signalling network behind the systemic resistance induced by Trichoderma harzianum in tomato. Front Plant Sci 4(6):206. https://doi.org/10.3389/fpls.2013.00206

162. Alizadeh H, Behboudi K, Ahmadzadeh M, Javan-Nikkhah M, Zamioudis C, Pieterse CMJ, Bakker PAHM (2013) Induced systemic resistance in cucumber and Arabidopsis thaliana by the combination of Trichoderma harzianum Tr6 and pseudomonas sp. Ps14. Biol Control 65(1):14–23. https://doi.org/10.1016/j.biocontrol.2013.01.009

163. Pozo MJ, Azcón-Aguilar C (2007) Unraveling mycorrhiza-induced resistance. Curr Opin Plant Biol 10(4):393–398. https://doi.org/10.1016/j.pbi.2007.05.004

164. Pozo MJ, Jung SC, López-Ráez JA, Azcón-Aguilar C (2010) Impact of arbuscular mycorrhizal symbiosis on plant response to biotic stress: the role of plant defence mechanisms. In: Koltai H, Kapulnik Y (eds) Arbuscular mycorrhizas: physiology and function. Springer, Dordrecht, pp 193–207. https://doi.org/10.1007/978-90-481-9489-6

165. Jung SC, Martínez-Medina A, López-Ráez JA, Pozo MJ (2012) Mycorrhiza-induced resistance and priming of plant defenses. J Chem Ecol 38(6):651–664. https://doi.org/10.1007/s10886-012-0134-6

166. Whipps JM (2004) Prospects and limitations for mycorrhizas in biocontrol of root pathogens. Can J Bot 82(8):1198–1227. https://doi.org/10.1139/b04-082

167. Gernns H, Alten H, Poehling H-M (2001) Arbuscular mycorrhiza increased the activity of a biotrophic leaf pathogen – is a compensation possible? Mycorrhiza 11(5):237–243. https://doi.org/10.1007/s005720100128

168. Liu J, Maldonado-Mendoza I, Lopez-Meyer-M, Cheung F, Town CD, Harrison MJ (2007) Arbuscular mycorrhizal symbiosis is accompanied by local and systemic alterations in gene expression and an increase in disease resistance in the shoots. Plant J 50 (3):529–544. https://doi.org/10.1111/j.1365-313X.2007.03069.x

169. Fritz M, Jakobsen I, Lyngkjær MF, Thordal-Christensen H, Pons-Kühnemann J (2006) Arbuscular mycorrhiza reduces susceptibility of tomato to Alternaria solani. Mycorrhiza 16 (6):413–419. https://doi.org/10.1007/s00572-006-0051-z

170. De La Noval B, Pérez E, Martínez B, León O, Martínez-Gallardo N, Délano-Frier J (2007) Exogenous systemin has a contrasting effect on disease resistance in mycorrhizal tomato (Solanum lycopersicum) plants infected with necrotrophic or hemibiotrophic pathogens. Mycorrhiza 17(5):449–460. https://doi.org/10.1007/s00572-007-0122-9

171. Hartley SE, Gange AC (2009) Impacts of plant symbiotic fungi on insect herbivores: mutualism in a multitrophic context. Annu Rev Entomol 54(1):323–342. https://doi.org/10.1146/annurev.ento.54.110807.090614

172. Pozo MJ, Cordier C, Dumas-Gaudot E, Gianinazzi S, Barea JM, Azcón-Aguilar C (2002) Localized versus systemic effect of arbuscular mycorrhizal fungi on defence responses to Phytophthora infection in tomato plants. J Exp Bot 53(368):525–534. https://doi.org/10.1093/jexbot/53.368.525

173. Gange AC (2007) Insect-mycorrhizal interactions: patterns, processes, and consequences. In: Ohgushi T, Craig TP, Price PW (eds) Ecological communities: plant mediation in indirect interaction webs. Cambridge University Press, Cambridge, pp 124–144. https://doi.org/10.1017/CBO9780511542701.007

174. Campos-Soriano L, García-Martínez J, Segundo BS (2012) The arbuscular mycorrhizal symbiosis promotes the systemic induction of regulatory defence-related genes in rice leaves and confers resistance to pathogen infection. Mol Plant Pathol 13(6):579–592. https://doi.org/10.1111/j.1364-3703.2011.00773.x

175. Song Y, Chen D, Lu K, Sun Z, Zeng R (2015) Enhanced tomato disease resistance primed by arbuscular mycorrhizal fungus. Front Plant Sci 6(9):786. https://doi.org/10.3389/fpls.2015.00786

176. Nair A, Kolet SP, Thulasiram HV, Bhargava S (2015) Systemic jasmonic acid modulation in mycorrhizal tomato plants and its role in induced resistance against Alternaria alternata. Plant Biol 17(3):625–631. https://doi.org/10.1111/plb.12277

177. Cordier C, Pozo MJ, Barea JM, Gianinazzi S, Gianinazzi-Pearson V (1998) Cell defense responses associated with localized and systemic resistance to Phytophthora parasitica induced in tomato by an Arbuscular Mycorrhizal fungus. Mol Plant-Microbe Interact 11 (10):1017–1028. https://doi.org/10.1094/MPMI.1998.11.10.1017

178. Garmendia I, Goicoechea N, Aguirreolea J (2004) Effectiveness of three Glomus species in protecting pepper (Capsicum annuum L.) against verticillium wilt. Biol Control 31 (3):296–305. https://doi.org/10.1016/j.biocontrol.2004.04.015
179. Kobra N, Jalil K, Youbert G (2009) Effects of three Glomus species as biocontrol agents against verticillium-induced wilt in cotton. J Plant Protect Res 49(2):185–189. https://doi.org/10.2478/v10045-009-0027-z
180. Sikes BA, Cottenie K, Klironomos JN (2009) Plant and fungal identity determines pathogen protection of plant roots by arbuscular mycorrhizas. J Ecol 97(6):1274–1280. https://doi.org/10.1111/j.1365-2745.2009.01557.x
181. Bernaola L, Cosme M, Schneider RW, Stout M (2018) Belowground inoculation with Arbuscular Mycorrhizal fungi increases local and systemic susceptibility of rice plants to different Pest organisms. Front Plant Sci 9 (6):747. https://doi.org/10.3389/fpls.2018.00747
182. Pozo MJ, Verhage A, García-Andrade J, García JM, Azcón-Aguilar C (2009) Priming plant defence against pathogens by Arbuscular Mycorrhizal fungi. In: Azcón-Aguilar C, Barea JM, Gianinazzi S, Gianinazzi-Pearson V (eds) Mycorrhizas- functional processes and ecological impact. Springer, Berlin, Heidelberg, pp 123–135. https://doi.org/10.1007/978-90-481-9489-6_9
183. Adolfsson L, Nziengui H, Abreu IN, Šimura J, Beebo A, Herdean A et al (2017) Enhanced secondary- and hormone metabolism in leaves of Arbuscular Mycorrhizal Medicago truncatula. Plant Physiol 175 (9):392–411. https://doi.org/10.1104/pp.16.01509
184. Holland JN, Cheng W, Crossley DA (1996) Herbivore-induced changes in plant carbon allocation: assessment of below-ground C fluxes using carbon-14. Oecologia 107 (1):87–94. https://doi.org/10.1007/BF00582238
185. Schultz JC, Appel HM, Ferrieri AP, Arnold TM (2013) Flexible resource allocation during plant defense responses. Front Plant Sci 4 (8):324. https://doi.org/10.3389/fpls.2013.00324

Chapter 5

Plant–Insect Bioassay for Testing *Arabidopsis* Resistance to the Generalist Herbivore *Spodoptera littoralis*

Stefan Mielke and Debora Gasperini

Abstract

Jasmonates are essential engineers of plant defense responses against many pests, including herbivorous insects. Herbivory induces the production of jasmonic acid (JA) and its bioactive conjugate jasmonoyl-L-isoleucine (JA-Ile), which then triggers a large transcriptional reprogramming to promote plant acclimation. The contribution of the JA pathway, including its components and regulators, to defense responses against insect herbivory can be evaluated by conducting bioassays with a wide range of host plants and insect pests. Here, we describe a detailed and reproducible protocol for testing feeding behavior of the generalist herbivore *Spodoptera littoralis* on the model plant *Arabidopsis thaliana* and hence infer the contribution of JA-mediated plant defense responses to a chewing insect.

Key words Plant-insect bioassay, Generalist herbivore, *Spodoptera littoralis*, Caterpillar, *Arabidopsis thaliana*, Jasmonic acid

1 Introduction

Plants growing in their natural environment are often attacked by a multitude of pests (e.g., insects, nematodes, bacteria, fungi, oomycetes, viruses) that exploit them as a source of nutrients or for reproductive purposes. To combat against invading pests, plants have evolved inducible defense mechanisms such as the production of specialized metabolites and proteins that exert toxic, repellent, or antinutritive effects to deter the attackers [1]. These responses are generally under hormonal control. Specifically, tissue damage and cellular insults provoked by herbivorous insects typically trigger an increase in the phytohormone jasmonoyl-L-isoleucine (JA-Ile), which, in turn, induces the transcriptional changes necessary to initiate plant defense responses [2, 3]. The crucial role of the JA pathway in mediating resistance against chewing insects was first demonstrated in the early 1990s [4, 5], and JA-mediated defense mechanisms have been intensely studied ever since. Nevertheless, in spite of plants' abilities to induce defense responses, herbivorous

Antony Champion and Laurent Laplaze (eds.), *Jasmonate in Plant Biology: Methods and Protocols*, Methods in Molecular Biology, vol. 2085, https://doi.org/10.1007/978-1-0716-0142-6_5,

insects still represent one of the major pests that destroy more than 20% of the net primary productivity in agriculture [6, 7]. Therefore, deciphering the mechanisms of JA-dependent plant–insect interactions could aid breeding strategies for eventually improving crop productivity.

Plant-insect bioassays have been often employed to determine the contribution of a given plant genotype to herbivore resistance [8]. They offer easily measurable phenotypes that address insect performance (insect size, weight, and survival rate) and can be correlated to expression profiles of defense-related genes or the amount of produced defense metabolites [9]. However, there is a plethora of insect herbivores that are affected by JA-induced defense responses and several ways to conduct a bioassay [10–13]. For instance, there are leaf-eating insects such as beetles and caterpillars [14, 15] as well as piercing-sucking herbivores, such as thrips and spider mites [16, 17] that can be assayed in plant host multiple-choice or no-choice bioassays [18]. Furthermore, some insect herbivores have dietary preferences and feed on a restricted set of plant species (specialist herbivores), while others endure on a variety of plant species (generalist herbivores). Specialists often coevolved with their host plants and have acquired effective countermeasures to overcome some plant defense responses, which will affect the bioassay's interpretation [19–22]. Hence, it may be necessary for researchers to test several bioassay systems, each requiring considerable optimization, before obtaining accurate data.

Here, we provide a detailed and reproducible method for determining the effects of plant defense responses to insect herbivory by using larvae of the Lepidopteran *Spodoptera littoralis* and the model plant *Arabidopsis thaliana*. Due to its wide host range, the generalist herbivore *S. littoralis* is a very destructive insect pest [23], to which wild-type (wt) *Arabidopsis* plants can partly defend against, whereas mutants in JA signaling (e.g., *coronatine insensitive1-1*, *coi1-1*) or JA biosynthesis (e.g., *allene oxide synthase*, *aos*) cannot [13, 24]. As the JA pathway plays a predominant role in mediating defense responses against chewing insects, and *Arabidopsis* JA mutants are readily available as controls, this combination provides a robust bioassay to test plant resistance of various plant genotypes. In fact, insect-feeding experiments using *S. littoralis* and *Arabidopsis* have been employed successfully in several studies, for example [25–30], with various procedures and expertise from individual laboratories. However, comparable and reproducible results from bioassays can only be obtained by ensuring a standardized setup and thoroughly executed experiments, as a multitude of factors may influence this biological system. Hence, this protocol is especially geared towards increasing reproducibility among individual experiments and across different labs. Additionally, the method may be used to screen mutants in putative novel regulators of the JA pathway, perform downstream assays on the challenged plants (e.g., transcriptomics, proteomics, metabolomics), and may be further adapted to test herbivore performance in other plant-insect combinations.

2 Materials

2.1 *Arabidopsis thaliana* Cultivation

1. Growth cabinet or growth chamber set at 21 °C, 60% relative humidity, light intensity of 100 μE/m^2/s, and short-day photoperiod (8 h light/16 h dark).
2. Heat-inactivated and cooled potting soil for *Arabidopsis* cultivation (*see* **Note 1**).
3. Pots for plant cultivation with perforated bottom (e.g., round pots with a diameter of 6.5 cm and total volume of 100 mL).
4. Plant trays (38 × 24 × 7 cm) fitting 15 pots, and respective transparent cover lids.
5. Tap water and watering can.
6. *Arabidopsis thaliana* wt seeds (e.g., Col-0) and a respective positive control, such as the JA-deficient *aos* mutant [31] (*see* **Note 2**).
7. Toothpick and a white piece of paper (e.g., 10 × 10 cm).
8. Cold room set at 4 °C for seed stratification.

2.2 *Spodoptera littoralis* Hatching

1. Low-form glass beaker (1 L, 108 mm diameter).
2. Whatman blotting paper 0.34 mm thick, cut into a 5 × 5 cm piece.
3. Autoclaved water.
4. Adhesive tape.
5. *S. littoralis* eggs on paper, from in-house or external supply.
6. Cling film (cut into 10 × 10 cm squares).
7. Parafilm M (cut into 10 × 10 cm squares).
8. Incubators set at 10 and 28 °C for insect hatching.

2.3 *Spodoptera littoralis–Arabidopsis* bioassay

1. 5-week-old *Arabidopsis* plants.
2. First-instar (L1) larvae of *S. littoralis*.
3. Bioassay cages (20 × 30 × 20 cm) made of 0.5 cm Plexiglas, with transparent closable lids and with holes (10 cm in diameter) on two opposite sides. Cover lid edges are lined with polyurethane foam and can be hooked with four rubber bands at the corners of each cage (Fig. 1a, *see* **Note 3**).
4. Nylon net with 100 μm pore size cut into 12 × 12 cm squares (*see* **Note 4**).

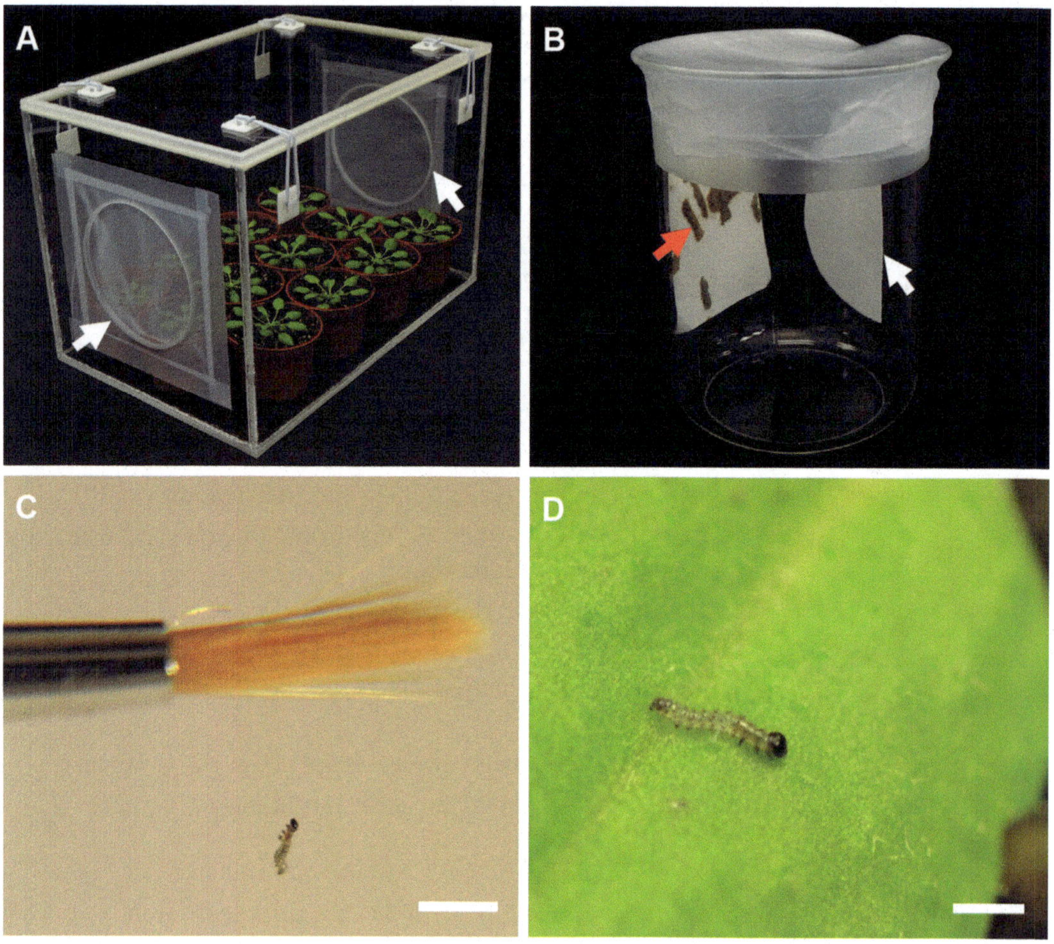

Fig. 1 Setup of *S. littoralis–Arabidopsis* bioassays. (**A**) A custom-made bioassay cage fitting 12 × 5-week-old plants grown in individual pots. Features of the cage include a removable transparent lid for easy transfer of plants and subsequent closure and two holes lined with a nylon net permitting gas exchange (white arrows). (**B**) A glass beaker for *S. littoralis* egg incubation and hatching. Moist Whatman paper (white arrow) and *S. littoralis* eggs on paper (red arrow) are taped inside the beaker, which is then sealed with cling film, Parafilm, and tape. (**C**) Freshly hatched *S. littoralis* larvae are placed onto rosette leaves by using a fine paintbrush to transfer the larvae by their silk. Scale bar = 3 mm. (**D**) Close-up of a freshly hatched *S. littoralis* larvae that was just transferred onto an *Arabidopsis* leaf. Scale bar = 1 mm

5. Fine paintbrush (size 2).
6. Fine blunt forceps (e.g., size 5).
7. Liquid nitrogen.
8. High-precision balance (0.00001 g accuracy).
9. Weighing dishes.
10. −20 °C freezer.
11. Camera for photographing.
12. Autoclave.

13. 70% (v/v) Ethanol.
14. Liquid dish soap.

3 Methods

3.1 Arabidopsis Cultivation

1. Fill labeled pots with soil, arrange them in trays, and water thoroughly with tap water from a watering can. Drain excess water from the trays once the soil is moist throughout the pots. For each genotype, prepare about 50% more pots than needed to ensure a sufficient number of uniformly grown plants (e.g., 18 pots per genotype).
2. Place a few *Arabidopsis* seeds on a piece of white paper and transfer one seed to the center of each pot with a moistened toothpick.
3. Cover plant trays with a transparent lid and stratify them for 2 days at 4 °C in the dark.
4. Transfer plant trays to the growth chamber and grow for 5 weeks (*see* **Note 5**).
5. Remove transparent lid covers after approximately 4 days or when cotyledons have opened.
6. Check plants daily and water them as needed (*see* **Note 6**).

3.2 Spodoptera littoralis Hatching

1. Wet a small piece of Whatman paper with autoclaved water and tape it inside the glass beaker (Fig. 1b). The paper should be moist but not dripping to ensure adequate humidity in the beaker during insect hatching.
2. Attach one or two pieces of paper containing *S. littoralis* eggs to the inside of a glass beaker using adhesive tape.
3. Close the beaker with one layer of cling film, fix it with a second layer of Parafilm, and then wrap the overhangs on the outside of the beaker with adhesive tape (*see* **Note 7**; Fig. 1b).
4. Incubate the beaker containing the insect eggs for 48 h at 10 °C in the dark (*see* **Note 8**). Afterward, transfer the beaker to a 28 °C incubator for 48 h to induce hatching.

3.3 Spodoptera littoralis: Arabidopsis bioassay

1. Plant age must be coordinated with insect hatching. To reduce variability, ensure that *S. littoralis* larvae are ready when plants are exactly 5 weeks old.
2. The day before starting the bioassay, water plants thoroughly to have adequate moisture for the following 10 days.
3. Organize the workspace for the bioassay. Cover the work surface with white paper to readily identify possible escaping larvae.

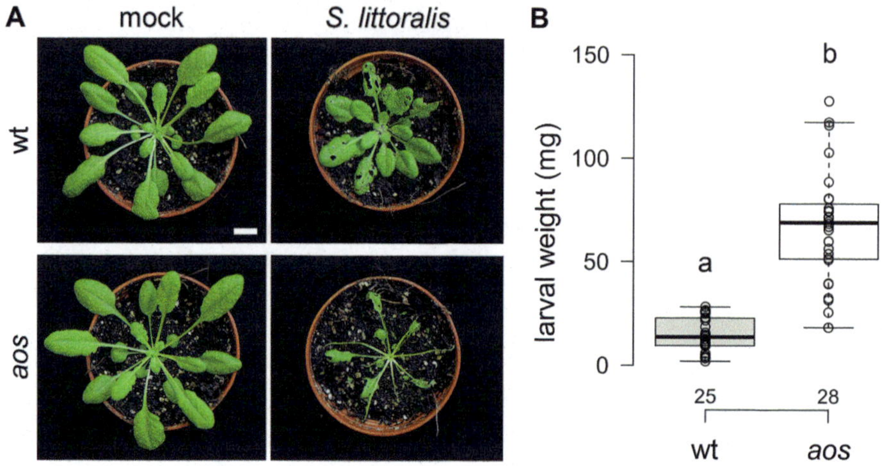

Fig. 2 Contribution of the JA pathway to plant resistance against the insect herbivore *S. littoralis*. (**A**) Rosette phenotypes of wt and JA-deficient *aos Arabidopsis* plants after 10 days of mock or challenge by *S. littoralis*. Scale bar = 1 cm. (**B**) Box plot summary of *S. littoralis* larval weights after feeding for 10 days on plants of indicated genotypes. Medians are represented by a solid line inside the box. Circles depict individual data points, and numbers at the bottom of each box plot represent *n*. Letters denote statistically significant differences as determined by Student's *t*-test (p-value = 1.55E^{-12})

4. For each genotype, select 12 uniform-looking 5-week-old plants.
5. Put the beaker with freshly hatched *S. littoralis* larvae into a flat white tray, gently tap the beaker so that larvae fall to the bottom, and then carefully remove the sealing film layers.
6. Next, carefully incline and tap the beaker, which will result in healthy larvae producing silk and abseiling.
7. Carefully transfer three insects to three different leaves of each plant, by picking larvae by the silk with a paintbrush (*see* **Note 9**; Fig. 1c, d).
8. Place pots with the larvae into the bioassay cage, for a total of 12 plants and 36 insects per cage (Fig. 1a).
9. Close bioassay cages and transfer them to the same growth cabinet or growth chamber where plants have been cultivated. Ensure a quiet environment and avoid manipulating the cages for the entire duration of the bioassay (*see* **Note 10**).
10. Allow larvae to feed undisturbed for 10 days, or until the positive control is consumed (Fig. 2a).
11. Once the bioassay is completed, transfer cages to a clean work-space covered with white paper. Carefully collect caterpillars from individual pots with blunt forceps and transfer the insects in weighing trays placed in liquid nitrogen to avoid weight loss caused by regurgitation. Inspect the abaxial side (underside) of each leaf, the topsoil, the entire pot surface, as well as the cage

to retrieve hiding insects. Caterpillars from individual bioassay cages consisting of a single-plant genotype are collected together in the same frozen tray.

12. Take photographs of a representative plant of each genotype that has been subjected to the bioassay. As a negative control, choose plants that underwent the assay without insect challenge (Fig. 2a). Plant tissues can be collected at this stage for further analysis.
13. Autoclave all pots and plant materials subjected to the bioassay to destroy possible remaining insects. Wash bioassay cages with soap and water and rinse them with 70% (v/v) Ethanol.
14. Weigh individual frozen caterpillars on a high-precision balance and record their weight. Summarize caterpillar's weights from individual plant genotypes in box plots and test differences with appropriate statistical tests (Fig. 2b). For example, when comparing more than two genotypes, use a one-way analysis of variance (ANOVA) followed by Tukey's honestly significant difference (HSD) test.
15. Given the high variability of this biological system, it is recommended to perform the experiment at least three times, using independent batches of cultivated plants and hatched *S. littoralis* larvae to verify reproducibility of newly assayed mutants.

4 Notes

1. We use the following soil mixture: 75% of Einheitserde Classic Kokos (Balster Einheitserdewerk, Fröndenberg, Germany, composed of 45% [w/w] white peat, 20% [w/w] clay, 15% [w/w] block peat, 20% [w/w] coco fibers) and 25% (w/w) of Vermiculite [Gärtnereibedarf Kamlott, Erfurt, Germany]). It is recommended to use fresh heat-inactivated or autoclaved soil, as potential contaminations from older soil (e.g., fungi, black flies) could affect the bioassay.
2. A JA-insensitive mutant such as *coi1-1* [32] can also be used as a positive control for the bioassay.
3. In our setup, the custom-made bioassay cages fit up to 12 pots. Cages of different sizes can be also used, as long as sufficient replication is ensured (e.g., ten plant pots per cage).
4. The nylon net is fixed with adhesive tape on the cage's inner sides to guarantee gas exchange and to prevent the escape of insects.
5. In our experience, most genotypes display an optimal size for bioassays after 5 weeks of growth under our specified conditions. Leaves should have expanded to the edges of the pot

without protruding. If slower-growing mutants are tested with respect to the controls, it may be feasible to grow them longer than 5 weeks, in order for plants to be homogenous. However, growth should be synchronized in order to start the bioassay with genotypes of equivalent developmental stages.

6. Ensure plants are kept at a uniform and constant moisture level. Avoid drought or overwatering, as this could affect plant growth and, ultimately, bioassay results.
7. The first layer of cling film is the most important, as it has to seal the beaker tightly in order to prevent hatched larvae from escaping. Always use a single piece of film to span the whole opening of the beaker. Sticking several pieces of film together for sealing is not recommended, as larvae might crawl between layers.
8. The 48-h egg incubation at 10 °C is the optimal time for uniform hatching. We have tested longer incubations at 10 °C (3, 4, 5, 6, and 8 days) and noticed that after 5 days, most eggs become unviable by turning color from brown to gray. Additionally, longer incubation times promote the formation of mold, which further reduces insect viability.
9. It is important to place larvae on different leaves to avoid cannibalism and to pick them by the silk instead of directly touching them to avoid any harm to the insects. It is also imperative not to touch, squeeze, or wound plant rosettes before and during the bioassay.
10. Vibrations might cause falling of young larvae to the bottom of the cage with consequent starvation and death. Hence, a stable environment should be guaranteed during the course of the bioassay.

Acknowledgments

We are grateful to O. Kindler and R. Reist (Stein, CH) for providing *S. littoralis* eggs; to E. Warkus (Halle, DE) for manufacturing the custom-made bioassay cages; and to the Deutsche Forschungsgemeinschaft (grant GA 2419/2-1 to D.G.) and IPB–Leibniz Association for funding.

References

1. Zhu-Salzman K, Luthe DS, Felton GW (2008) Arthropod-inducible proteins: broad spectrum defenses against multiple herbivores. Plant Physiol 146(3):852–858. https://doi.org/10.1104/pp.107.112177
2. Fonseca S, Chini A, Hamberg M, Adie B, Porzel A, Kramell R, Miersch O, Wasternack C, Solano R (2009) (+)-7-iso-Jasmonoyl-L-isoleucine is the endogenous bioactive jasmonate. Nat Chem Biol 5(5):344–350. https://doi.org/10.1038/nchembio.161

3. Koo AJ, Howe GA (2009) The wound hormone jasmonate. Phytochemistry 70 (13–14):1571–1580. https://doi.org/10.1016/j.phytochem.2009.07.018
4. Farmer EE, Johnson RR, Ryan CA (1992) Regulation of expression of proteinase inhibitor genes by methyl jasmonate and jasmonic acid. Plant Physiol 98(3):995–1002
5. Farmer EE, Ryan CA (1990) Interplant communication: airborne methyl jasmonate induces synthesis of proteinase inhibitors in plant leaves. Proc Natl Acad Sci U S A 87 (19):7713–7716
6. Agrawal AA (2011) Current trends in the evolutionary ecology of plant defence. Functional Ecology 25(2):420–432. https://doi.org/10.1111/j.1365-2435.2010.01796.x
7. Gatehouse JA (2002) Plant resistance towards insect herbivores: a dynamic interaction. New Phytologist 156(2):145–169. https://doi.org/10.1046/j.1469-8137.2002.00519.x
8. Schuman MC, Baldwin IT (2016) The layers of plant responses to insect herbivores. Annu Rev Entomol 61:373–394. https://doi.org/10.1146/annurev-ento-010715-023851
9. Verhage A, Vlaardingerbroek I, Raaymakers C, Van Dam NM, Dicke M, Van Wees SC, Pieterse CM (2011) Rewiring of the jasmonate signaling pathway in Arabidopsis during insect herbivory. Front Plant Sci 2:47. https://doi.org/10.3389/fpls.2011.00047
10. Howe GA, Jander G (2008) Plant immunity to insect herbivores. Annu Rev Plant Biol 59:41–66. https://doi.org/10.1146/annurev.arplant.59.032607.092825
11. McConn M, Creelman RA, Bell E, Mullet JE, Browse J (1997) Jasmonate is essential for insect defense in Arabidopsis. Proc Natl Acad Sci U S A 94(10):5473–5477
12. Moran PJ, Thompson GA (2001) Molecular responses to aphid feeding in Arabidopsis in relation to plant defense pathways. Plant Physiol 125(2):1074–1085
13. Stotz HU, Koch T, Biedermann A, Weniger K, Boland W, Mitchell-Olds T (2002) Evidence for regulation of resistance in Arabidopsis to Egyptian cotton worm by salicylic and jasmonic acid signaling pathways. Planta 214 (4):648–652
14. Dombrecht B, Xue GP, Sprague SJ, Kirkegaard JA, Ross JJ, Reid JB, Fitt GP, Sewelam N, Schenk PM, Manners JM, Kazan K (2007) MYC2 differentially modulates diverse jasmonate-dependent functions in Arabidopsis. Plant Cell 19(7):2225–2245. https://doi.org/10.1105/tpc.106.048017
15. Kessler A, Halitschke R, Baldwin IT (2004) Silencing the jasmonate cascade: induced plant defenses and insect populations. Science 305(5684):665–668. https://doi.org/10.1126/science.1096931
16. Abe H, Onnishi J, Narusaka M, Seo S, Narusaka Y, Tsuda S, Kobayashi M (2008) Arabidopsis-thrips system for analysis of plant response to insect feeding. Plant Signal Behav 3 (7):446–447
17. Zhurov V, Navarro M, Bruinsma KA, Arbona V, Santamaria ME, Cazaux M, Wybouw N, Osborne EJ, Ens C, Rioja C, Vermeirssen V, Rubio-Somoza I, Krishna P, Diaz I, Schmid M, Gomez-Cadenas A, Van de Peer Y, Grbic M, Clark RM, Van Leeuwen T, Grbic V (2014) Reciprocal responses in the interaction between Arabidopsis and the cell-content-feeding chelicerate herbivore spider mite. Plant Physiol 164(1):384–399. https://doi.org/10.1104/pp.113.231555
18. Knolhoff LM, Heckel DG (2014) Behavioral assays for studies of host plant choice and adaptation in herbivorous insects. Annu Rev Entomol 59:263–278. https://doi.org/10.1146/annurev-ento-011613-161945
19. Ali JG, Agrawal AA (2012) Specialist versus generalist insect herbivores and plant defense. Trends Plant Sci 17(5):293–302. https://doi.org/10.1016/j.tplants.2012.02.006
20. Gols R, Bukovinszky T, van Dam NM, Dicke M, Bullock JM, Harvey JA (2008) Performance of generalist and specialist herbivores and their endoparasitoids differs on cultivated and wild Brassica populations. J Chem Ecol 34 (2):132–143. https://doi.org/10.1007/s10886-008-9429-z
21. Kliebenstein D, Pedersen D, Barker B, Mitchell-Olds T (2002) Comparative analysis of quantitative trait loci controlling glucosinolates, myrosinase and insect resistance in Arabidopsis thaliana. Genetics 161(1):325–332
22. Lankau RA (2007) Specialist and generalist herbivores exert opposing selection on a chemical defense. New Phytol 175(1):176–184. https://doi.org/10.1111/j.1469-8137.2007.02090.x
23. Health EPP (2015) Scientific Opinion on the pest categorisation of Spodoptera littoralis. EFSA J 13:1. https://doi.org/10.2903/j.efsa.2015.3987
24. Yan Y, Stolz S, Chetelat A, Reymond P, Pagni M, Dubugnon L, Farmer EE (2007) A downstream mediator in the growth repression limb of the jasmonate pathway. Plant Cell 19 (8):2470–2483. https://doi.org/10.1105/tpc.107.050708

25. Gasperini D, Chetelat A, Acosta IF, Goossens J, Pauwels L, Goossens A, Dreos R, Alfonso E, Farmer EE (2015) Multilayered organization of jasmonate signalling in the regulation of root growth. PLoS Genetics 11(6): e1005300. https://doi.org/10.1371/journal.pgen.1005300
26. Kanchiswamy CN, Takahashi H, Quadro S, Maffei ME, Bossi S, Bertea C, Zebelo SA, Muroi A, Ishihama N, Yoshioka H, Boland W, Takabayashi J, Endo Y, Sawasaki T, Arimura G (2010) Regulation of Arabidopsis defense responses against Spodoptera littoralis by CPK-mediated calcium signaling. BMC Plant Biol 10:97. https://doi.org/10.1186/1471-2229-10-97
27. Khan GA, Vogiatzaki E, Glauser G, Poirier Y (2016) Phosphate deficiency induces the jasmonate pathway and enhances resistance to insect herbivory. Plant Physiol 171 (1):632–644. https://doi.org/10.1104/pp.16.00278
28. Reymond P, Bodenhausen N, Van Poecke RM, Krishnamurthy V, Dicke M, Farmer EE (2004) A conserved transcript pattern in response to a specialist and a generalist herbivore. Plant Cell 16(11):3132–3147. https://doi.org/10.1105/tpc.104.026120
29. Scholz SS, Vadassery J, Heyer M, Reichelt M, Bender KW, Snedden WA, Boland W, Mithofer A (2014) Mutation of the Arabidopsis calmodulin-like protein CML37 deregulates the jasmonate pathway and enhances susceptibility to herbivory. Mol Plant 7 (12):1712–1726. https://doi.org/10.1093/mp/ssu102
30. Schweizer F, Fernandez-Calvo P, Zander M, Diez-Diaz M, Fonseca S, Glauser G, Lewsey MG, Ecker JR, Solano R, Reymond P (2013) Arabidopsis basic helix-loop-helix transcription factors MYC2, MYC3, and MYC4 regulate glucosinolate biosynthesis, insect performance, and feeding behavior. Plant Cell 25 (8):3117–3132. https://doi.org/10.1105/tpc.113.115139
31. Park JH, Halitschke R, Kim HB, Baldwin IT, Feldmann KA, Feyereisen R (2002) A knock-out mutation in allene oxide synthase results in male sterility and defective wound signal transduction in Arabidopsis due to a block in jasmonic acid biosynthesis. Plant J 31(1):1–12
32. Xie DX, Feys BF, James S, Nieto-Rostro M, Turner JG (1998) COI1: an Arabidopsis gene required for jasmonate-regulated defense and fertility. Science 280(5366):1091–1094

Chapter 6

Targeted Metabolomics of Plant Hormones and Redox Metabolites in Stomatal Immunity

Lisa David, Jianing Kang, and Sixue Chen

Abstract

Phytohormones and redox metabolites are important molecules in a number of biological processes related to plant growth, development, and stress responses. Understanding how these metabolites are involved in abiotic and biotic stress is a frequent topic of plant biology research. However, many factors, such as low physiological concentrations and the inherent complexity of plant samples, make identification and quantification of these important metabolites difficult. Here, we describe a method for metabolite extraction from whole leaves and guard cell–enriched samples and a targeted metabolomics strategy for the identification and quantification of specific hormone- and redox-related metabolites. In our experiment, we used the reference plant *Arabidopsis thaliana* infected with the biotrophic pathogen *Pseudomonas syringe* pv. tomato (*Pst*) DC3000, and examined the changes in hormone and redox metabolites in systemic leaves, using the targeted metabolomics strategy in order to investigate potential functions of these metabolites in systemic acquired resistance (SAR) during a plant's immune responses. The methods reported here can be expanded to other metabolites and other biological systems beyond plants and bacterial pathogens.

Key words Systemic acquired resistance, Phytohormones, Redox metabolites, Targeted metabolomics, *Arabidopsis thaliana*, *Pseudomonas syringae*

1 Introduction

Phytohormones play critical roles in responses to a multitude of abiotic and biotic stresses, including drought, salinity, cold, wounding, herbivory, and pathogens [1]. The cross talk in elaborate hormone signaling networks allows plants to perceive simultaneous environmental signals and adapt to adverse conditions to maintain optimal growth and development [1, 2]. The hormones abscisic acid (ABA), salicylic acid (SA), jasmonic acid (JA) and jasmonates, and ethylene (ET) have long been known to play major roles in plant defense responses to pathogens and abiotic stresses [1, 2]. Generally, ABA levels increase in plant response to abiotic stresses like drought and salinity [3]. On the other hand, JA, SA, and ET increase in responses to biotic stresses such as herbivory and

Antony Champion and Laurent Laplaze (eds.), *Jasmonate in Plant Biology: Methods and Protocols*, Methods in Molecular Biology, vol. 2085, https://doi.org/10.1007/978-1-0716-0142-6_6, © Springer Science+Business Media, LLC, part of Springer Nature 2020

pathogen infections [1, 2]. Recently, responses of these hormones in plant defense have been shown to include cross talk with auxins, gibberellins, and cytokinins [1, 2]. Upon perceiving abiotic and biotic stresses, several complex signaling pathways are initiated in plants, which include hormone signaling, alteration of intracellular Ca^{2+} concentrations, production of second messengers such as reactive oxygen species (ROS) and inositol phosphate, and activation of kinase cascades [1, 2, 4–6]. In concert with Ca^{2+} increases, a rapid increase in ROS production, referred to as "oxidative burst," often occurs in response to various stress conditions [4]. In *Arabidopsis thaliana*, ROS molecules such as hydrogen peroxide (H_2O_2) activate mitogen-activated protein kinase (MAPK) cascades and upregulate transcription of stress-related genes [1, 2, 5, 6].

Systemic acquired resistance (SAR) is an induced immune mechanism in plants, in which previous exposure by one part of the plant leads to a broad-spectrum immune response in the uninfected parts of the plant [7]. An avirulent pathogen, such as *Pseudomonas syringae* pv. tomato (*Pst*) DC3000, can induce SAR through generation of mobile signals, accumulation of the defense hormone SA, and secretion of the antimicrobial PR (pathogenesis-related) proteins [8–12]. As a result of SAR signaling, the rest of the plant is protected from secondary infection for a period of weeks to months [7]. SAR signaling also involves the generation of ROS as important signaling components in the regulation of defense responses [4]. Cellular redox changes associated with ROS and oxidative stress in plant defense have been well documented [4, 7]. ROSs are formed in the apoplast by plant nicotinamide adenine dinucleotide phosphate (NADPH) oxidases, but their accumulation has also been observed in chloroplasts, mitochondria, and even nuclei [4].

Guard cells are specialized cells that in pairs form the stomata. They respond to many different environmental cues and control the opening and closing of the stomatal pores. Research has shown that guard cells are pathogen responsive, in that they close the stomatal pores and therefore play an important role in plant immunity [13–17]. Perception of pathogen-associated molecular patterns (PAMPs) by guard cell pattern recognition receptors (PRRs) triggers calcium channel activation in guard cells, increasing Ca^{2+}, which then activates calcium-dependent protein kinases (CDPKs) [15, 16]. CDPKs then activate plasma membrane–localized NADPH oxidases (RBOHs) and induce extracellular ROS production [15, 16]. Stomatal closure has been shown by numerous studies to be affected by the phytohormones SA, JA, and ABA [13, 15, 16, 18].

The study of plant hormone and redox dynamics is complicated, first, by the complexity of the stress response networks, and, second, by the often-minute physiological concentrations of the metabolites [19]. Metabolomics, a study of a complete set of metabolites present in a specific cell or organism, offers a method to identify and quantify these important molecules. Targeted metabolomics allows researchers to direct the sensitive and selective analysis to a set of targeted metabolites in biological samples. Targeted metabolomics platforms derive information from discovery experiments and literature and use as a model or hypothesis, which can be tested experimentally in a quantitative manner. This testing involves validation of the metabolites from large sample sets and distinguishing closely related compounds at low concentration levels. A popular targeted metabolomics approach, selected reaction monitoring (SRM), or multiple reaction monitoring (MRM), performed on a triple quadrupole mass spectrometer, allows for sensitive, high-throughput, and reliable quantitation of selected metabolites.

To investigate the roles of hormones and redox metabolites in whole leaves and guard cells, we used MRM-targeted metabolomics to quantify a specific set of hormone and redox metabolites (Table 1, Fig. 1) in systemic tissues of *Arabidopsis* plants that had been previously infected with the pathogenic bacteria *Pst* DC3000 to induce SAR. We collected systemic leaves, as well as guard cell–enriched epidermal peels from systemic leaves of *Arabidopsis* plants after infection with bacteria in a time-course experiment to determine the changes in specific hormone and redox-regulated metabolites in response to SAR. We utilized a TSQ Altis triple quadrupole system (Thermo Scientific Inc.) for MRM analysis of the metabolite compounds (Table 1). The tandem mass analyzers of the TSQ Altis scan a precursor and product ion (transition) pair, providing both high-sensitivity and high-throughput analysis of samples (Fig. 1). Hormone metabolites included ABA, gibberellic acid (GA) 3, indole-3-acetic acid (IAA), SA, JA, methyljasmonic acid (MeJA), JA-isoleucine conjugate (JA-Ile), and 12-oxo-phytodienoic acid (OPDA) (Fig. 2). Redox metabolites included ascorbate, dehydroascorbate, cysteine, cystine, glutathione (GSH), oxidized glutathione (GSSG), methionine, and methionine sulfoxide (Fig. 3).

2 Materials

2.1 Plant Growth and Maintenance

1. *Arabidopsis thaliana* ecotype Columbia seeds obtained from the Arabidopsis Biological Resource Centre (ABRC).
2. Soil-Metromix MVP, The Scotts Co., Marysville, OH, USA.
3. Growth chamber maintained at 22 °C (light/dark) and an 8 h/16 h (light/dark) photoperiod.
4. Incandescent bulbs capable of emitting 150 μmol m^{-2} s^{-2}.

Table 1
Mass spectrometer settings for hormone and redox metabolite analysis

Metabolite	Polarity	Precursor (*m/z*)	Product (*m/z*)	Collision energy (V)	Dwell time (ms)	RF lens (V)
Hormones						
Abscisic acid	Negative	263.03	153.13	10.27	13.55	50.00
(ABA)	Negative	263.03	219.15	12.62	13.55	50.00
Gibberellin 3	Positive	347.09	311.09	10.68	25.81	90.03
(GA3)	Positive	347.09	329.08	11.02	25.81	90.03
Gibberellin 4	Negative	331.30	243.17	18.83	13.55	81.00
(GA4)	Negative	331.30	287.24	17.13	13.55	81.00
Indole-3-acetic acid	Negative	174.00	128.07	19.63	13.55	32.00
(IAA)	Negative	174.00	130.05	10.23	13.55	32.00
Indole-3-carboxylic acid	Positive	162.01	102.96	16.03	25.81	86.11
(ICA)	Positive	162.01	143.99	10.23	25.81	86.11
Jasmonic acid	Negative	209.30	59.00	11.44	13.55	48.00
(JA)	Negative	209.30	165.13	12.81	13.55	48.00
Jasmonic acid-isoleucine	Negative	322.80	130.13	20.31	25.81	81.00
(JA-Ile)	Negative	322.80	131.08	20.69	25.81	81.00
Methyljasmonic acid	Positive	225.20	133.05	15.95	13.55	41.00
(MeJA)	Positive	225.20	151.16	12.92	13.55	41.00
12-Oxo-phytodienoic acid	Negative	291.20	165.16	19.14	13.55	83.00
(12-OPDA)	Negative	291.20	247.22	18.11	13.55	83.00
Traumatic acid	Negative	227.00	165.14	10.23	13.55	49.00
(TA)	Negative	227.00	183.14	13.15	13.55	49.00
Zeatin	Positive	220.12	136.00	18.00	25.84	57.74
	Positive	220.12	202.13	13.34	25.84	57.74
Redox metabolites						
Ascorbate	Negative	175.01	87.04	18.53	11.15	51.00
	Negative	175.01	114.97	12.16	11.15	51.00
Dehydroascorbate	Negative	172.99	71.01	11.44	11.15	43.00
	Negative	172.99	142.97	10.23	11.15	43.00
Cysteine	Positive	122.07	56.00	14.55	11.15	32.00

(continued)

Table 1 (continued)

Metabolite	Polarity	Precursor (*m/z*)	Product (*m/z*)	Collision energy (V)	Dwell time (ms)	RF lens (V)
	Positive	122.07	76.00	14.28	11.15	32.00
Cystine	Positive	241.03	73.97	27.33	11.15	43.00
	Positive	241.03	151.98	13.23	11.15	43.00
Glutathione	Positive	308.09	162.04	16.33	11.15	51.00
(GSH)	Positive	308.09	179.07	11.67	11.15	51.00
Glutathione oxidized	Positive	613.26	354.95	21.64	11.15	131.00
(GSSG)	Positive	613.26	483.99	15.61	11.15	131.00
Methionine	Positive	150.04	56.00	17.00	11.15	30.00
	Positive	150.04	132.91	10.00	11.15	30.00
Methionine sulfoxide	Positive	166.11	74.00	13.00	11.15	47.00
	Positive	166.11	148.83	11.00	11.15	47.00
Spiked-in standards						
Camphorsulfonic acid	Negative	231.10	79.92	41.73	25.84	92.00
	Negative	231.10	149.05	25.96	25.84	92.00
Lidocaine	Positive	235.26	58.07	31.00	17.25	64.00
	Positive	235.26	86.08	21.00	17.25	64.00

The top precursor/product transition for each metabolite is used for quantitation, and the bottom is for qualification. Also listed are the collision energies used for fragmentation of each metabolite, the dwell time in milliseconds for each metabolite, as well as the radio frequency (RF) lens used to guide the ions in the quadrupole analyzer

2.2 Bacterial Growth and Preparation

1. *Pseudomonas syringae* pv. tomato strain (*Pst*) DC3000.
2. Laminar flow hood.
3. King's B media: 1.5-g anhydrous K_2HPO_4, 15-g glycerol, 5 mL $MgSO_4$ (1 M; sterile), 15-g Agar, Antibiotics (25-mg rifampicin, 50-mg kanamycin), 1L H_2O, Adjust the pH to 7.0 with HCl, Autoclave.
4. 10 mM $MgCl_2$.
5. Round petri plates.
6. Centrifuge.
7. Needleless syringe.

2.3 Preparation of Guard Cell–Enriched Epidermal Peels

1. High-speed blender.
2. Ice.
3. Liquid nitrogen.
4. Razor blades.

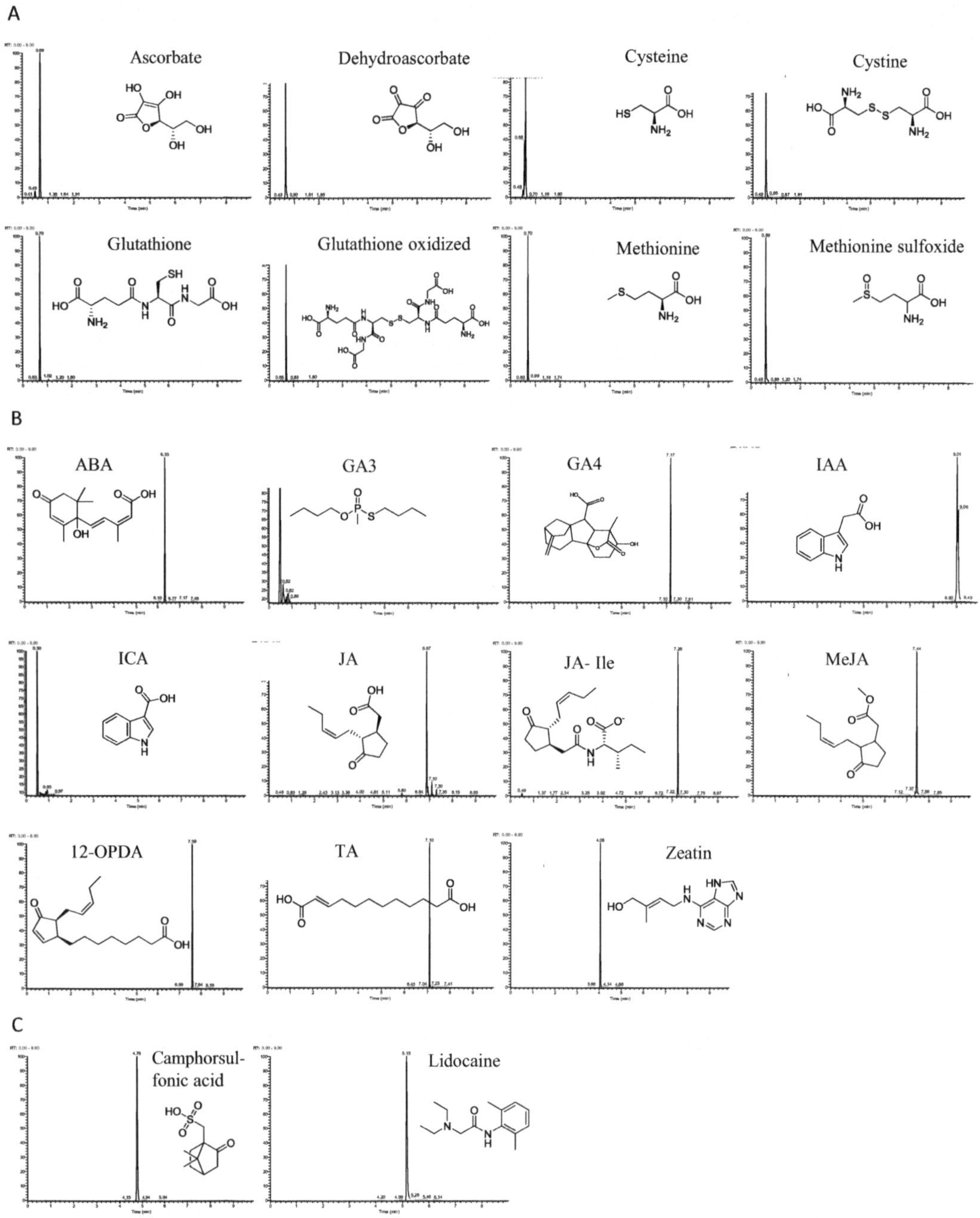

Fig. 1 MRM chromatograms of redox and hormone metabolites. Retention times and the MRM transitions in Table 1 for the metabolite standards were used to identify distinct peaks from different hormone and redox metabolites in plant samples. (**a**) Peaks and chemical structures of redox metabolites including ascorbate, dehydroascorbate, cysteine, cystine, glutathione (GSH), glutathione oxidized (GSSG), methionine, and methionine sulfoxide. (**b**) Peaks and chemical structures for hormone metabolites including abscisic acid (ABA), gibberellic acid (GA) 3, GA4, indole-3-acetic acid (IAA), indole-3-carboxylic acid (ICA), jasmonic acid (JA), JA-isoleucine conjugate (JA-Ile), methyljasmonic acid (MeJA), 12-oxo-phytodienoic acid (OPDA), traumatic acid (TA), and salicylic acid (SA). (**c**) Peaks and chemical structures of the internal standards used, including lidocaine for positive mode and (1S)-(+)-10-camphorsulfonic acid for negative mode

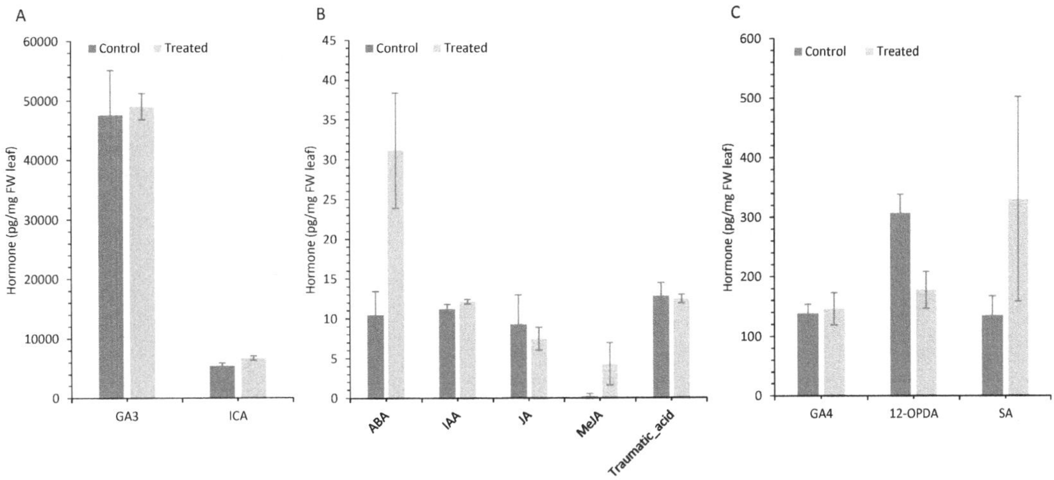

Fig. 2 Changes of hormone metabolites in guard cell–enriched epidermal peels after systemic acquired resistance (SAR). Hormone metabolites from noninjected systemic leaves were quantified from guard cell–enriched epidermal peel samples in SAR (treated) and mock (control) plants at 72 h after initial injection with either *Pst* or mock (10 mM $MgCl_2$). For each sample, 15 epidermal peels from 3 plants of noninjected leaves were collected from systemic tissue of 5-week-old *Arabidopsis* plants grown in 8/16 h light/dark cycle and 22°C. The average and standard deviation of four biological replicates of each sample were calculated. (**a**) GA3 and ICA. (**b**) ABA, IAA, JA, MeJA, and TA. (**c**) GA4, 12-OPDA, and SA (refer to Table 1 and Fig. 1 legends for abbreviations)

5. 200-μm nylon mesh filter.
6. 15-mL Falcon tubes for storage.

2.4 Metabolite Extraction from Whole Leaf and Epidermal Peels

1. Internal standards: Lidocaine for positive mode, (1S)-(+)-10-Camphorsulfonic acid for negative mode. Internal standard stock mix: 0.1 nmol/μL of each compound dissolved in ddH_2O.
2. Extraction buffer I: Acetonitrile: Isopropanol: H_2O = 3:3:2 (v: v:v).
3. Extraction buffer II: Acetonitrile: H_2O = 1:1 (v:v).
4. Vortexer Genie 2 manufactured by Scientific Industries. Catalog number 00-SI-0236
5. Formic acid

2.5 Mass Spectrometry and Data Analysis

1. Vanquish ultra-high-performance liquid chromatography system.
2. Hypersil GOLD™ aQ column (100 mm length × 2.1 mm diameter, 1.9 μm particle size, Thermo Scientific, Waltham, Massachusetts, USA).
3. HPLC Solvent A: 0.1% formic acid.
4. HPLC Solvent B: 0.1% formic acid, 99.9% acetylnitrile (ACN).

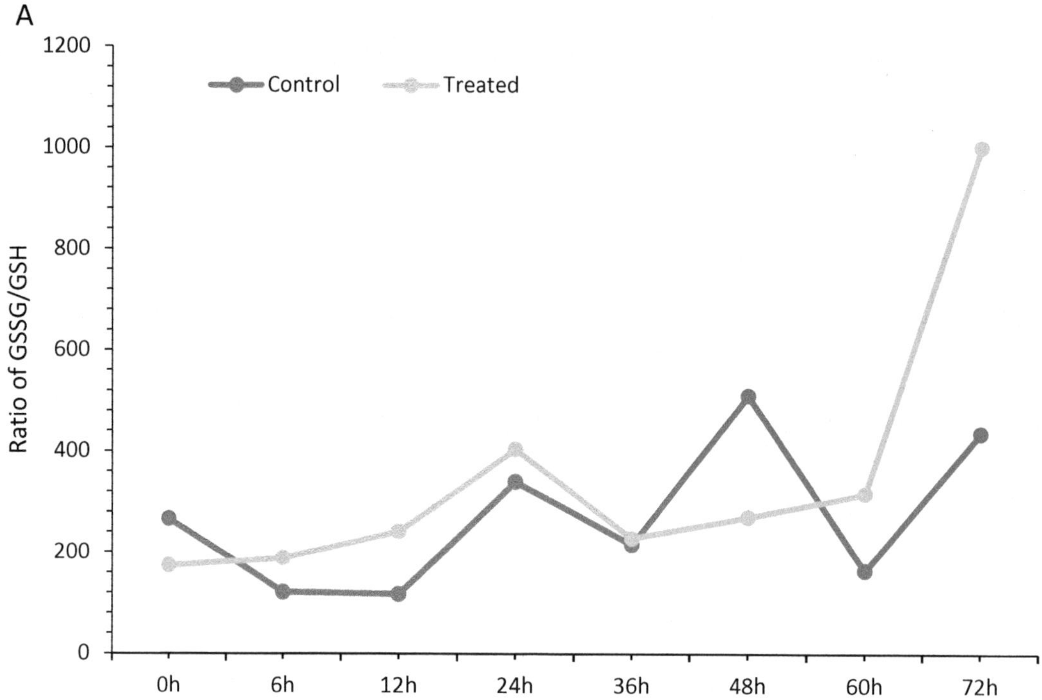

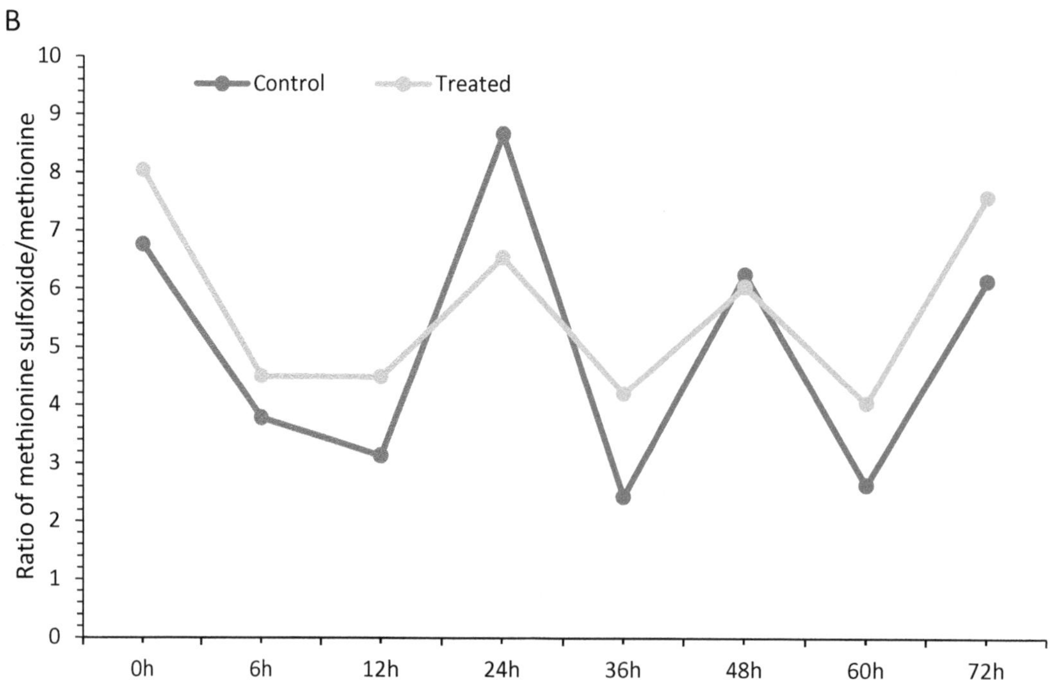

Fig. 3 Temporal changes of example redox metabolites in the development of SAR in leaves. Redox metabolites were quantified from whole leaf samples in SAR (treated) and mock (control) plants over a time course from 0 to 72 h. A total of 100 mg fresh weight of noninjected leaves were collected at 0, 6, 12, 24, 36, 48, 60, and 72 h after initial injection with either *Pst* or mock (10 mM $MgCl_2$) from systemic tissue of 5-week-old *Arabidopsis* plants grown in 8/16 h light/dark cycle and 22 °C. The ratios of pairs of oxidized: reduced metabolites were calculated from the averages of four biological replicates. (**a**) Ratio of GSSG:GSH. (**b**) Ratio of methionine sulfoxide:methionine

5. TSQ Altis triple quadrupole mass spectrometer (Thermo Scientific).
6. Xcalibur software (Thermo Scientific).
7. Microsoft Excel.

3 Methods

3.1 Plant Growth and Maintenance

1. After cold treatment at 4 °C in the dark for 2 days in ddH_2O, *A. thaliana* ecotype Columbia seeds were allowed to germinate on pre-moistened soil in a growth chamber under high humidity by placing a transparent dome over the growth tray. The seedlings were transferred individually into fresh pre-moistened soil and were grown for a total of five weeks in the same growth chamber under a light intensity from incandescent bulbs of 150 $\mu mol \cdot m^{-2}\ s^{-2}$, 22 °C, and an 8-h light/16-h dark photoperiod cycle (*see* **Notes 1** and **2**).
2. For whole leaf samples: collect 0.5 g fresh weight leaf tissue in aluminum foil. Freeze leaves immediately in liquid nitrogen and store in −80 °C for further analyses (*see* **Note 3**).
3. For guard cell–enriched epidermal peel samples, follow steps in Subheading 3. Freeze epidermal peels immediately in liquid nitrogen and store collection in −80 °C until further analyses.

3.2 Bacterial Growth, Preparation, and Infiltration into Leaves

1. Culture the strain *Pst* DC3000 on King's B agar plates supplemented with 50 mg/L kanamycin and 25 mg/L rifampicin by streaking the plate and incubating for 18 h in 25 °C.
2. Grow a liquid culture by taking a single colony from the agar plate and inoculating liquid King's B medium for 18 h in 25 °C until the $OD_{600} = 0.6$ (*see* **Note 4**).
3. Harvest bacterial cells by centrifugation at 10,000 rpm (10,621 × *g*) for 1 min in a microcentrifuge.
4. Resuspend bacterial cells in 10 mM $MgCl_2$ to $OD_{600} = 0.02$ and use for infection of plants.
5. Use a needleless syringe to infiltrate the rosette leaves of *A. thaliana* from the abaxial surface with *Pst* DC3000 as described by Huang et al. [5] (*see* **Note 5**).
6. Dry underside of leaf with a Kimwipe to remove any excess bacterial solution.
7. Leave the plants to grow under specified growth conditions (*see* Subheading 3.1) until collection of systemic (non-infected) tissue.

3.3 Preparation of Guard Cell–Enriched Epidermal Peels

This method was adapted from Bauer et al. [20].

1. Collect 4 mature leaves/plant that are not injected with the bacteria from 36 Arabidopsis plants, for a total of 144 leaves in each sample replicate.
2. Remove the central vein of each leaf with a razor blade and place immediately in ice water.
3. After collection of 144 leaves, place the sample in a high-speed blender with ice and 250 mL of cold Milli-Q water.
4. After blending for 1 min, pour the sample through the 200-μm nylon mesh (*see* **Note 6**).
5. Rinse the mesh into the blender cup to collect the tissue that did not pass through the mesh.
6. Repeat this process of blending and filtering two more times for a total of three times to collect the final guard cell–enriched sample.
7. After the third blending and collection, dry the underside of the mesh on a paper towel briefly, then scrape the sample off the mesh and into a 15-mL Falcon tube.
8. Immediately freeze the sample in liquid nitrogen and store in −80 °C.

3.4 Metabolite Extraction from Whole Leaf and Epidermal Peels

1. Weigh out 0.5 g of either whole leaf or guard cell–enriched peel samples.
2. Add 1 mL extraction buffer I and 10 μL internal standard stock to sample.
3. Vortex at 1100 rpm (vortex setting #5 on the Vortexer Genie 2), 4 °C for 15 min.
4. Sonicate for 15 min.
5. Centrifuge at 13,000 rpm (17,949 × *g*) in a microcentrifuge, 4 °C for 15 min.
6. Remove the supernatant.
7. Repeat extraction steps once more with 1 mL of extraction buffer II.
8. Combine the supernatants and lyophilize extracts.
9. Resuspend extracts with 100 μL of ddH_2O and 0.1% formic acid.
10. Centrifuge at 13,000 rpm (17,949 × *g*) in a microcentrifuge, 4 °C for 15 min.
11. Transfer the supernatant to autosampler vials for HPLC-MRM. Be careful not to take any debris (*see* **Note 7**).
12. For statistical confidence, obtain at least four independent replicates of each sample.

3.5 Mass Spectrometry and Data Analysis

1. Transfer at least 10 μL of metabolite sample in 0.1% formic acid to HPLC autosampler vials.
2. Set the flow rate of the HPLC to 0.45 mL/min. Set the gradient as follows: 0–2 min 0% B, 2–4 min 5% B, 4–6 min 20% B, 6–7 min 98% B, 7–8 min 98% B, 8–8.5 min 0% B, 8.5–10 min 0% B (*see* **Notes 8** and **9**).
3. Set the parameters of the MS scan in the TSQ Altis triple quadrupole system as follows: Resolution: 0.7 (*see* **Notes 10** and **11**).
4. Inject 10 μL of samples into the HPLC and then acquire the MS1 and MS2 spectra.
5. Use the Thermo Xcalibur software to verify the peaks of metabolites in the targeted scan and quantify the area of the peaks (*see* **Note 12**).
6. Export all data into an Excel file.
7. Perform appropriate statistical analyses such as a Student *t*-test to examine differential quantities of selected metabolites in the two test groups of samples (control vs. treated).

4 Notes

1. During growth and maintenance of the *Arabidopsis* plants, it is very important that they do not suffer from other sources of environmental stress, as your purpose is to study the effects of your imposed biological stress. This requires that if you germinate the seeds on soil in groups, they must be transplanted before they reach 3 weeks of age, and this must be done carefully so as not to damage the roots or shoots of the seedlings. They need to be given enough space to develop in appropriate soil. And they must be watered appropriately for their stage of development. Overwatering or under-watering your plants during their development will impose environmental stress responses in the tissues that you will collect for analyses. One indicator of environmental stress in the seedlings is production of anthocyanins in the leaf tissue. When you collect the leaf tissue, it should appear green in color and turgid. If the leaves have a dark or purple color or seem soft, dry, or papery to the touch, they have been affected by environmental stresses during their growth period.
2. The reason we grow the *Arabidopsis* plants in 8-h light/16-h dark photoperiod cycle for 5 weeks is that we have found this increases the amount of vegetative tissue (leaves) and delays the flowering time of the plants. If you are using a 16 h- or 24 h-light cycle, the collection of the leaves may need to be earlier than 5 weeks to ensure that the vegetative tissue is viable, and

you may need to collect tissue from more leaves as the leaves would probably be smaller.

3. Collection of leaf tissue should be done as quickly as possible with minimal damage to the leaves. Collect the tissue from the live plant while you are next to the weigh balance and use the aluminum foil as the weigh paper. After weighing out the leaf tissue, immediately wrap the tissue in the aluminum foil and freeze in liquid nitrogen.
4. The reason for growing the bacteria overnight in a liquid culture is that it is important for the bacteria to be in the log phase of their reproduction when you use them to infiltrate the leaves of the plants. This ensures that you are injecting active bacteria cells into your plant tissue.
5. When using the needleless syringe to infiltrate the leaves, try to fill the majority of the leaf with the bacterial solution. Lightly press the adaxial (upper) side of the leaf to your forefinger, while placing the needleless syringe against the abaxial (lower) side of the leaf. Slowly depress the syringe to allow the solution to fill the leaf. Do not overfill or press too hard with the syringe on the leaf. If you overfill, you may damage the edges of the leaf, and pressing hard against the leaf can cause wounding.
6. When pouring the sample through the 200-μm nylon mesh, it helps to secure the mesh. In Bauer et al, 2013, the authors secured the nylon mesh inside an embroidery loop. We have used the top of a plastic bottle like a funnel and secured the mesh to the opening of the bottle with rubber bands while we pour the sample through it from the blender cup.
7. Transferring the supernatant to the HPLC vials without getting debris in the sample can be difficult, but it is important to avoid debris as much as possible so that the sample will not clog the HPLC system. To achieve this, we have transferred 90 μL of the total 100 μL sample after centrifugation, and then we have repeated the centrifugation step a second time and only 50 μL of the 90 μL sample is transferred to the HPLC vial after the second centrifugation. For HPLC, you will only need about 10 μL of sample for each injection, and you can do multiple injections as technical replicates.
8. During the optimization step, in order to make the target peak stable and sharp, you must adjust parameters such as flow rate, sheath, aux, sweep, and CID gas to make the spray stable between 5 and 15% relative standard deviation (RSD). Also, between each injection with different samples, the syringe and sample transfer line should be flushed with isopropanol. When performing mass spectral scanning of the sample, a blank (0.1% formic acid, 99.9% ACN) needs to be added between biological replicates to avoid carryover between different samples.

9. Transfer at least 10 μL of standard samples of different concentrations in 0.1% formic acid to HPLC autosampler vials. Set flow rate of HPLC to 0.45 mL/min and set the gradient as follows:0–2 min 0% B, 2–4 min 5% B, 4–6 min 20% B, 6–7 min 98% B, 7–8 min 98% B, 8–8.5 min 0% B, 8.5–10 min 0% B. Based on the transitions, collision energies, and retention time (RT) on the column of the previously obtained standards, finalize the method and run the samples.
10. Due to the intrinsic complexity of the plant tissues, when the detected peak in the data analysis was performed and compared with the standard, the actual plant metabolite RT sometimes shifted slightly, had higher noise, higher baseline, and double peaks. Thus, it is necessary to optimize some parameters such as baseline window and number of scans over which to look for local minima. The area noise factor is a noise-level multiplier used to determine a peak's edge after locating a possible peak. Peak noise factor is a noise-level multiplier used to determine potential peak signal threshold, and increasing this parameter is useful to select the correct peak in the case of a double peak. Sometimes, it is important to spike in the authentic standards to ensure the correct peak identification.
11. Optimization: 500 ng/mL MRM standard solution is prepared for further optimization. Use a microsyringe in combination with autosampler to inject (flow rate: 30 μL/mL) the standard solution into the TSQ Altis MS. Optimization is based on the chemical formula of the standard. Select the top 3–4 in terms of intensity of the transition and determine the collision energy of the transition based on the change in intensity.
12. Use the Xcalibur processing program to verify the peak area of metabolites by characteristic RT and transitions. The peak area of the metabolite is brought into the standard curve of the metabolite standards to obtain its quantitative value.

References

1. Verma V, Ravindran P, Kumar PP (2016) Plant hormone-mediated regulation of stress responses. BMC Plant Biol 16:86. https://doi.org/10.1186/s12870-016-0771-y
2. Nguyen D, Rieu I, Mariani C, van Dam NM (2016) How plants handle multiple stresses: hormonal interactions underlying responses to abiotic stress and insect herbivory. Plant Mol Biol 91:727–740. https://doi.org/10.1007/s11103-016-0481-8
3. Fan J, Hill L, Crooks C, Doerner P, Lamb C (2009) Abscisic acid has a key role in modulating diverse plant-pathogen interactions. Plant Physiology 150:1750–1761. https://doi.org/10.1104/pp.109.137943
4. Van Breusegem F, Bailey-Serres J, Mittler R (2008) Unraveling the tapestry of networks involving reactive oxygen species in plants. Plant Physiol 147:978–984. https://doi.org/10.1104/pp.108.122325
5. Misra BB, Chaturvedi R (2015) When plants brace for the emerging pathogens. Physiol Mol Plant Pathol 92:181–185. https://doi.org/10.1016/j.pmpp.2015.03.004
6. Spoel SH, Dong XN (2012) How do plants achieve immunity? Defense without specialized

immune cells. Nat Rev Immunol 12 (2):89–100. https://doi.org/10.1038/nri3141

7. Fu ZQ, Dong XN (2013) Systemic acquired resistance: turning local infection into global defense. Annu Rev Plant Biol 64:839–863. https://doi.org/10.1146/annurev-arplant-042811-105606
8. Park SW, Kaimoyo E, Kumar D, Mosher S, Klessig DF (2007) Methyl salicylate is a critical mobile signal for plant systemic acquired resistance. Science 318(5847):113–116. https://doi.org/10.1126/science.1147113
9. Kachroo A, Robin GP (2013) Systemic signaling during plant defense. Curr Opin Plant Biol 16:527–533. https://doi.org/10.1016/j.pbi.2013.06.019
10. Sels J, Mathys J, De Coninck BMA, Cammue BPA, De Bolle MFC (2008) Plant pathogenesis-related (PR) proteins: a focus on PR peptides. Plant Physiol Biochem 46 (11):941–950. https://doi.org/10.1016/j.plaphy.2008.06.011
11. Shah J, Zeier J (2013) Long-distance communication and signal amplification in systemic acquired resistance. Front Plant Sci 4 (30):1–16. https://doi.org/10.3389/fpls.2013.00030
12. Dempsey DA, Klessig DF (2012) SOS – too many signals for systemic acquired resistance? Trends Plant Sci 17(9):538–545. https://doi.org/10.1016/j.tplants.2012.05.011
13. Zeng W, Brutus A, Kremer JM, Withers JC, Gao X, Jones AD et al (2011) A genetic screen reveals arabidopsis stomatal and/or apoplastic defenses against *Pseudomonas syringae* pv. *tomato* DC3000. PLoS Pathol 7:10. https://doi.org/10.1371/journal.ppat.1002291
14. Melotto M, Underwood W, He SY (2008) Role of stomata in plant innate immunity and foliar bacterial diseases. Annu Rev Phytopathol 46:101–122. https://doi.org/10.1146/annurev.phyto.121107.104959
15. McLachlan DH, Kopischke M, Robatzek S (2014) Gate control: guard cell regulation by microbial stress. New Phytol 203:1049–1063. https://doi.org/10.1111/nph.12916
16. Sawinski K, Mersmann S, Robatzek S, Bohmer M (2013) Guarding the green: pathways to stomatal immunity. Mol Plant-Microbe Interact 26:626–632. https://doi.org/10.1094/MPMI-12-12-0288-CR
17. Melotto M, Underwood W, Koczan J, Nomura K, He SY (2006) Plant stomata function in innate immunity against bacterial invasion. Cell 126(5):969–980. https://doi.org/10.1016/j.cell.2006.06.054
18. Assmann SM, Jegla T (2016) Guard cell sensory systems: recent insights on stomatal responses to light, abscisic acid, and CO_2. Curr Opin Plant Biol 33:157–167. https://doi.org/10.1016/j.pbi.2016.07.003
19. Shah J, Chaturvedi R, Chowdhury Z, Venables B, Petros RA (2014) Signaling by small metabolites in systemic acquired resistance. Plant J 79(4):645–658. https://doi.org/10.1111/tpj.12464
20. Bauer H, Ache P, Lautner S, Fromm J, Hartung W, Al-Rasheid KA, Sonnewald S, Sonnewald U, Kneitz S, Lachmann N, Mendel RR, Bittner F, Hetherington AM, Hedrich R (2013) The stomatal response to reduced relative humidity requires guard cell-autonomous ABA synthesis. Curr Biol 23:53–57. https://doi.org/10.1016/j.cub.2012.11.022

Chapter 7

Bioassays to Evaluate the Resistance of Whole Plants to the Herbivorous Insect Thrips

Merel Steenbergen, Colette Broekgaarden, Corné M. J. Pieterse, and Saskia C. M. Van Wees

Abstract

Thrips are tiny, cell-content–feeding insects that are a major pest on crops and ornamentals. Besides causing direct feeding damage, thrips may also cause indirect damage by vectoring tospoviruses. Novel resistance mechanisms to thrips need to be discovered and validated. Induction of jasmonic acid–dependent defenses has been demonstrated to be essential for resistance to thrips, but underlying mechanisms still need to be discovered. For this, it is vital to use robust plant-thrips assays to analyze plant defense responses and thrips performance. In recently developed high-throughput phenotyping platforms, the feeding damage that is visible as silver spots, and the preference of thrips in a two-choice setup is assessed, using leaf discs. Here, we describe whole-plant thrips assays that are essential for (1) validation of findings obtained by the leaf disc assays, (2) assessment of longer-term effects on thrips feeding success and fecundity, (3) determination of spatial-temporal effects induced by primary thrips infestation on a secondary attack by thrips or other insects or pathogens, and (4) assessment of gene expression and metabolite changes. We present detailed methods and tips and tricks for (a) rearing and selection of thrips at different developmental stages, (b) treatment of the whole plant or an individual leaf with thrips, and (c) determination of feeding damage and visualization of thrips oviposition success in leaves.

Key words Cell-content–feeding insect, Thrips, *Arabidopsis thaliana*, Plant immunity, Jasmonate, Clip cage, Aspirator, Oviposition, Rearing

1 Introduction

Thrips are tiny insects (1.5 mm or less) that are cell-content feeders, greatly affecting commercial food crops and ornamentals worldwide. Besides the direct feeding damage that thrips cause, indirect damage can be inflicted by the *tospoviruses* that can be transmitted in a non-ingesting probe. The life cycle of thrips comprises six stages: the egg stage, two larvae stages (first instar L1 and second instar L2), two pupae stages (prepupa and pupa), and the adult stage (Fig. 1; reviewed by Ref. 1). Female thrips lay their bean-shaped eggs in the epidermal layer of various above-ground plant

Antony Champion and Laurent Laplaze (eds.), *Jasmonate in Plant Biology: Methods and Protocols*, Methods in Molecular Biology, vol. 2085, https://doi.org/10.1007/978-1-0716-0142-6_7,

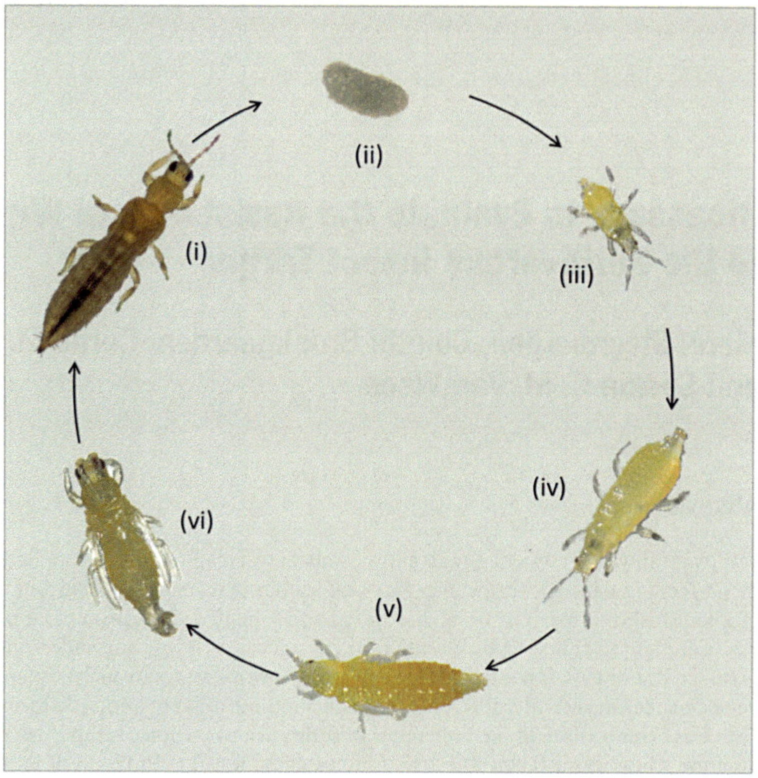

Fig. 1 Thrips life cycle. Adult female thrips (i) can lay fertilized or non-fertilized eggs (ii) from which a female or a male thrips larva will emerge, respectively, after 3–4 days (depending on the temperature, which is optimal at 27 °C). The first instar larva (L1; iii) will develop into a second instar larva (L2; iv) within 2 days. The L2 larva will go through a pre-pupa (v) and pupa (vi) phase after 2–3 days, from which an adult thrips emerges after 4–5 days

parts, like leaves, stems, and flowers, using a saw-like ovipositor [2, 3]. Fertilized (diploid) eggs will develop into female thrips, and unfertilized (haploid) eggs will give rise to male thrips. The larvae that hatch from the eggs will go through two larval stages that actively feed from the plant. During the two pupae stages, thrips are nonfeeding and move to the soil to prepare for the adult stage. Compared to the larvae, the adults are more mobile because of their fringed wings, allowing them to fly off and look for new suitable host plants (reviewed by Ref. 1). Both larval and adult thrips feed from the plant by piercing with their stylet in epidermal, mesophyll, and parenchymal cells and ingesting the contents [4, 5]. Emptied cells collapse or fill with air which, after intensive feeding, results in a localized silvery appearance (reviewed by Ref. 1).

Upon probing and ingestion of the plant cells, different defense responses are initiated in the plant. This activation of defenses requires the perception and recognition of specific and general herbivore- and damage-associated molecular patterns (HAMPs and DAMPs). Direct and/or indirect defenses against thrips are coordinated by phytohormones such as jasmonic acid (JA), salicylic acid (SA), abscisic acid (ABA), and ethylene (ET). JA has been demonstrated to be essential in the defense against many herbivore species, including thrips (reviewed by Ref. 1), for example, plants insensitive to JA or deficient in JA accumulation are more susceptible to thrips [6]. Subsequent to triggering of the JA pathway by thrips, different direct and indirect defense mechanisms are activated, like formation of leaf trichomes and production of phenolic compounds with toxic characteristics and volatile organic compounds that can attract natural predators of thrips. There is a growing body of evidence for the ability of herbivores to manipulate JA-associated defenses through the action of effector proteins in the arthropod saliva that are secreted into the plant cells, but whether thrips harbor these saliva-borne effectors is still unknown (reviewed by Ref. 1). Another form of defense manipulation that has not been described yet for thrips, but that has been demonstrated for chewing herbivores, is that via egg-associated cues that trigger SA signaling, which suppresses anti-herbivore defenses, thereby benefitting the future progeny of thrips [7].

Thrips are too tiny to measure their weight or length. Therefore, to determine the performance of thrips larvae and adults on different plants, the amount of silver damage inflicted by their feeding is often monitored. Another measure to determine the performance is the reproductive success of thrips on the plants. Moreover, if given a choice between different genotypes or treatments, the preference of thrips for certain plant material, if given a choice, can be assessed. Recently, two high-throughput phenotyping platforms have been developed to quantify host-plant resistance/thrips preference. The first one uses a two-choice setup of half-leaf discs in a 96-well format in which the behavior of the thrips is tracked continuously by a video camera throughout a period of several hours and analyzed with commercial software [8]. The second one is based on a no-choice setup of leaf discs in a petri dish in which the amount of silver damage inflicted by thrips at one endpoint is captured by a photo camera and quantified by freeware [9]. These two screening methods will aid to reduce the quantity of material, and the space and time needed to screen for thrips resistance compared to screening of whole plants. Comparisons between leaf disc and/or detached leaf tests with whole-plant tests revealed that the relatively high amount of damage inflicted to the leaf material prior to the start of a leaf disc assay did not influence the thrips resistance scores of a few selected pepper and *Arabidopsis thaliana* accessions [8–10].

Nonetheless, good setups for whole-plant performance assays are still essential. For example, to study spatial-temporal effects induced by primary thrips infestation on resistance to a secondary attack by thrips or other insects or by pathogens, whole-plant assays are required. Also, it is difficult to maintain leaf discs in good shape, and therefore, in order to study longer-term effects on the thrips performance in terms of feeding and oviposition success, the use of intact plants is essential. Moreover, findings based on leaf disc assays should be validated under greenhouse or field conditions with whole plants. And lastly, for sensitive gene expression or metabolite accumulation assays, it is preferred to introduce as little variation as possible, which demands for less handling and surely not damaging the tissue, before thrips are introduced. Here, we describe several methods for thrips assays on whole plants. We explain how to rear and select different developmental stages of thrips, for which different tools can be handmade. We present how to treat the whole plant or individual leaves with thrips, for bioassays in which feeding damage or oviposition success can be assessed, or for the study of induced defense responses, locally and systemically. The use of proper thrips–plant assays, as described here, will help to increase our understanding of the defense mechanisms underlying plant resistance to thrips.

2 Materials

2.1 Equipment Used for Multiple Purposes

1. 1.5 mL Eppendorf tubes.
2. Pipet and tips (200 μL, 1000 μL).
3. Fridge.
4. Vacuum autoclave (344 L) and autoclavable plastic bags (40 × 60 cm).
5. River sand.
6. Potting soil.
7. Modified half-strength Hoagland nutrient solution [11]: 5 mM KNO_3, 5 mM $Ca(NO_3)_2$, 2 mM KH_2PO_4, 2 mM $MgSO_4$, trace elements, pH 7, and 10 mM Fe-ethylenediamine-di[o-hydroxyphenylacetic acid] (Sequestreen).
8. Potting bench.
9. Plant labels.
10. Small trays (150–300 mL; 4 cm high).
11. Big trays (L45 × W30 × H8 cm) with transparent lids.
12. Garden felt.
13. Forceps (curved and straight tip).
14. 15-mL tubes.
15. 1.5-mL safe-lock tubes.
16. Small petri dishes (Ø 6 cm).
17. Scissors.

18. Cork-borer (Ø 4 mm, Ø 15 mm, Ø 20 mm).
19. Bunsen burner.
20. Matches.
21. Heat-resistant gloves.
22. Nylon mesh (80-micron).
23. Hot glue gun.
24. Liquid nitrogen.

2.2 Plant Cultivation

2.2.1 A. thaliana

1. Climate-controlled growth chambers with a 10-h day and 14-h night cycle set at 21 °C and 70% relative humidity. Light provided by HPI-T Plus lamps (400W/645 E40) (*see* **Note 1**).
2. Seeds of *A. thaliana*.
3. 0.1% (w/v) agar.
4. Pots (60 mL, Ø 5.5 cm, height 5 cm) with holes in the bottom for plant cultivation after the seedling stage.

2.2.2 Lettuce

1. A controlled greenhouse compartment with natural light conditions at 18 °C and 70% relative humidity.
2. Seeds of *Lactuca sativa var. capitate*.
3. Pots (800 mL, L9 × W9 × H10 cm) with holes in the bottom for plant cultivation after the seedling stage.
4. Plastic plant pot saucers (Ø 18 cm).

2.3 Instruments for Thrips Rearing and Assays

2.3.1 Containers for Synchronized Larval Rearing and Whole-Plant Bioassays

1. 1-L transparent polyethylene terephthalate (PET) jars with screw cap (Ø 10.4 cm).
2. Butyl septa (Ø 20 mm).
3. Hole saw (Ø 6 cm).

2.3.2 Clip Cages for Single Leaf Treatment with Thrips

1. Acrylate rings (Ø inside, 21 mm; thickness, 4 mm; height, 5 mm and 8 mm).
2. Metal hairclips.
3. 4-cm squares of polyether foam (thickness of 3 mm).
4. Two-component glue plastic.

2.3.3 Aspirator for Collecting Thrips of a Specific Life Stage

1. A 15-mL plastic, transparent pot with screw cap (Ø of 25 mm).
2. Polytetrafluorethylene tubes (Ø 4 mm, length 30 cm).

2.4 Thrips Rearing

1. A controlled growth cabinet with a 10-h day and 14-h night cycle at 26 °C, 70% relative humidity, and a light intensity of 100 μmol m^{-2} s^{-1}.
2. BugDorms (L60 × D60 × H60 cm) with 150-micron mesh.
3. Common bean pods (*Phaseolus vulgaris L.*) from the supermarket.
4. Bee pollen.
5. PCR tubes (200 μL) and caps.
6. Paper tissues.
7. Fine paintbrush.
8. White 500 mL polypropylene box (L182 × W135 × H35 mm) with transparent lid.

2.5 Trypan Blue Staining

1. Trypan blue solution (lactic acid, glycerol, demi-water, and 96% ethanol (1:1:1:3, v:v:v:v) + 0.025% trypan blue (adapted from Ref. 12).
2. Rack for 15-mL tubes.
3. Floating tube rack for Eppendorf tubes.
4. Water bath.
5. Pipette-boy.
6. 25-mL serological pipettes.
7. Empty glass bottle (1L).
8. Lactoglycerol solution (lactic acid, glycerol and demi-water (1:1:1, v:v:v)).
9. Petri dish (Ø 15 cm).
10. Microscope glass slide.
11. 95% ethanol.
12. Stereo zoom optical microscope (10× magnification).

3 Methods

3.1 Plant Cultivation

3.1.1 *A. thaliana* Cultivation

1. Suspend *A. thaliana* seeds in 1.0 mL of 0.1% agar in 1.5-mL Eppendorf tubes and store at 4 °C for 3 days.
2. Autoclave river sand in double autoclavable bags for 45 min at 120 °C.
3. Autoclave a mix of potting soil and river sand (12:5, v:v) in double autoclavable bags twice for 45 min at 120 °C with a 24-h interval.
4. Add half-strength Hoagland to sand (150 mL/kg sand). Fill up the 4-cm-high small trays with sand until 1 cm below the edge.

5. Distribute seeds (in 0.1% agar) evenly on sand using a 1000-μL pipet. Place the small tray in a big tray (L45 × W30 × H8), add some water to the bottom of the big tray, and cover with the transparent lid to ensure 100% relative humidity. Place in a growth chamber for 10–14 days.
6. Use a potting bench to mix Hoagland solution (50 mL/kg) and an equal amount of water with the sterile potting soil mixture.
7. Cut garden felt in pieces of L45 × W30 cm (the size of big tray), wet with water, and place in big plant trays (*see* **Note 2**).
8. Fill 60-mL pots with the soil mix, pots should weigh approximately 70 g, and make a hole of 1.5 cm deep in the soil of each pot with the conical end of a 15-mL tube. Place the pots in the big trays with felt.
9. To transfer a seedling to the planting hole, flood the small sand trays with water, and use forceps with a curved tip to loosen up the sand and to gently pull out the seedling.
10. After transfer of the seedling into the planting hole, close the hole gently by pushing the soil back around the root.
11. Stick a plant label (colored or with text) in each pot to indicate the genotype or treatment. Randomize the different genotypes/treatments in and between trays (40 pots/tray).
12. Cover the trays with transparent lids for 2 days to ensure 100% relative humidity. Crack the lids for 1 day before full removal to allow the plants to gradually acclimate to a relative humidity of 70% for the following weeks.
13. Every other day, water the plants until the felt is saturated. Once a week, apply 500 mL of Hoagland solution per tray.

3.1.2 Lettuce Cultivation

1. Follow **steps 1–6** of Subheading 3.1.1 for cultivation of lettuce seedlings.
2. Fill 800-mL pots with sterile potting soil (not a mixture with sand) and place them in a big tray (eight pots/tray; *see* **Note 3**). Make a hole of 3 cm deep in the soil of each pot with the conical end of a 15-mL tube.
3. Gently remove a lettuce seedling from sand using your fingers and place it in the planting hole. Close the hole gently by pushing the soil back around the root.
4. From here, follow **steps 10**, **12**, and **13** of Subheading 3.1.1 for cultivation of the lettuce plants after transfer of seedlings.

Fig. 2 Modified PET-container for thrips larval rearing and whole-plant bioassays

3.2 Tools for Rearing and Collecting of Thrips, and for Bioassays

3.2.1 Containers for Larval Rearing and Whole-Plant Bioassays

For both larval rearing and whole-plant bioassays, modified 1-L transparent PET containers (Ø 10.4 cm) with screw caps are used (Fig. 2).

1. Using a hot glue gun, apply some glue underneath the bottom of a 6-cm petri dish to attach it on the bottom inside the container.
2. Use a hole saw with a drill to make a Ø 6-cm hole in the white screw cap of the container.
3. Cut circles (Ø 8 cm) from 80-micron mesh and use a glue gun to attach the mesh onto the inside of the screw cap.
4. Heat the tip of a cork-borer (Ø 20 mm) and create a hole on the side of the container at 2 cm from the bottom. Cover the hole with a 20-mm butyl septum.

3.2.2 Clip Cages for Single Leaf Assays on Whole Plants

For assays in which a localized treatment with a defined number of thrips on one leaf is needed, a clip cage can be used, which allows the treated leaf to remain attached to an intact plant (Fig. 3). Clip-cage assays are not only suitable for gene expression or metabolite studies but also to determine systemic effects of thrips feeding on secondary infestation or infection. By selecting leaves of a similar developmental stage, variation between replicates will be reduced.

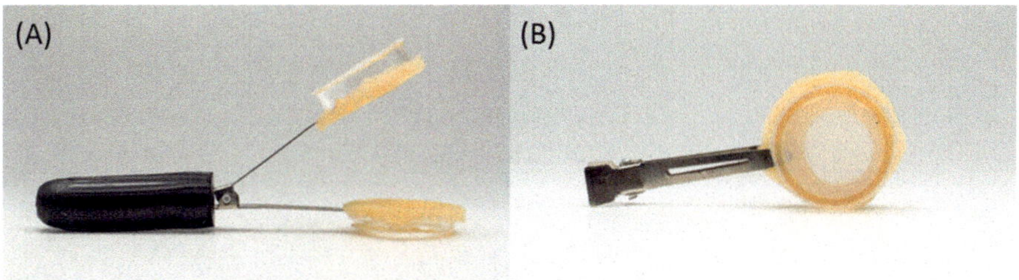

Fig. 3 Clip cage. (**a**) Side view of the clip cage when opened. (**b**) Top/bottom view of the clip cage. Foam is glued on the inside and mesh on the outside of the rings

Clip cages can not only be ordered online but can also be made in the lab as described here.

1. Hold two acrylate rings, one of 5 mm height and one of 8 mm height, between the thumb and the index finger.
2. Use a Bunsen burner to heat the tips of the metal hair clips. Wear heat-resistant gloves when doing this. When sufficiently heated (the tip will start to glow), each tip of the clip (bottom or top) can be gently pressed into the side of the rings (one tip in the top ring, the other in the bottom one).
3. Cut 80-micron mesh into squares of 4 × 4 cm and using a glue gun attach one mesh on the outside of the 5-mm ring and one on the outside of the 8-mm ring.
4. Cut the polyether foam into squares of 4 × 4 cm and make stacks of 10.
5. Wear heat-resistant gloves when heating the tip of the Ø 15-mm cork-borer in the burner. Press it through the middle of the foam stack to create a hole in each patch.
6. Glue one of the foam patches on the inside of the 5-mm ring and another one on the inside of the 8-mm ring (*see* **Note 4**).
7. When the glue is dry, cut the edges of the mesh and the foam.

3.2.3 Aspirator for Adult Thrips Collection

An aspirator is used to collect female thrips. Adult females are used for the whole-plant bioassays because then both feeding damage and oviposition can be assessed as performance parameters. Females are selected from a colony reared in the BugDorm (*see* Subheading 3.3) and can be recognized by their dark abdomen. Male thrips are smaller and have a lighter color (Fig. 4). Aspirators can be bought (online) but can also easily be handmade (Fig. 5), following the protocol below.

1. Heat the tip of a Ø 4-mm cork-borer in a Bunsen burner and pierce two holes in the screw cap of a 15-mL plastic, transparent pot.

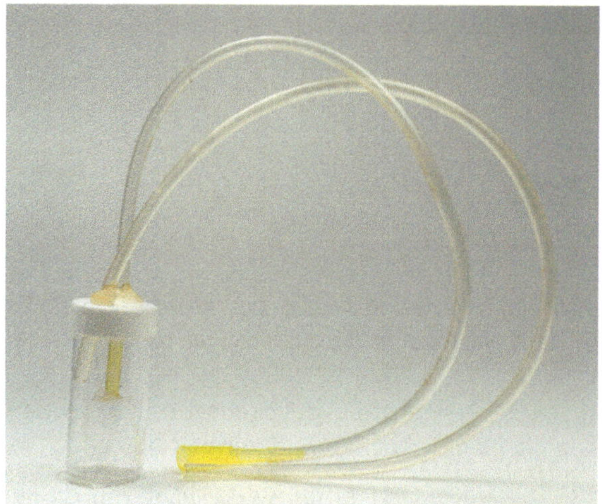

Fig. 4 Handmade aspirator to collect thrips

Fig. 5 Adult male (left) and female (right) thrips

2. Cut the silicone tubes into two pieces of 30-cm length. Slide the tubes through the holes of the screw cap (until 1 cm below the cap) and glue-shut the opening between the tube and the hole using a glue gun.
3. Cut 80-micron mesh into small squares of 5 × 5 mm and glue it to the end of one of the tubes, facing the inside of the pot.
4. Cut 1 cm off the tip of a 200-μL pipette tip and slide it into the other end of the tube that carries the mesh.

3.3 Thrips Rearing and Collecting

3.3.1 Adults

When starting a thrips colony, set up two to three BugDorms. When you start a colony with 100 adult thrips (mix of males and females), you can have 100–200 adult female thrips after approximately 18 days. To prevent inbreeding, mix the colonies between the BugDorms and import thrips from other thrips sources (e.g., from collaborators) every 6 months.

1. Place the BugDorm in a controlled growth cabinet at 26 °C, which is an ideal temperature for thrips.
2. Place four plastic plant pot saucers in each BugDorm. Cut pieces of garden felt (Ø 18 cm) to put in the saucers.
3. Place a pot with a 5-week-old lettuce plant on each saucer (*see* Subheading 3.1.2 for the cultivation of lettuce).
4. Add a mix of male and female thrips to each BugDorm.
5. Cover the bottom of a small petri dish with bee pollen and place the dish in the BugDorm (*see* **Note 5**). Refresh the bee pollen weekly.
6. Every other day, water the plants until the felt is saturated. Once a week, apply Hoagland solution (*see* Subheading 3.1.2).
7. For collection of female thrips, gently shake a thrips-infested lettuce plant from the BugDorm in the white 500-mL box and close it with a transparent lid.
8. Place the pipette tip of the aspirator made into Subheading 3.2.3 in your mouth; gently lift the corner of the lid of the box and cover the selected thrips with the end of the other tube to suck it up. The thrips is now collected in the pot, and the mesh at the end of the 'sucking' tube prevents the thrips from moving into that tube.
9. If the desired amount of thrips is collected (e.g., 5 for a leaf- or whole-plant assay and 100 for synchronized rearing of thrips larvae), then remove the aspirator cap and screw on a normal cap (without holes).

3.3.2 Larvae

After thrips larvae hatch from the eggs, they develop two larval stages. The transition from the first to the second larval stage is difficult to distinguish. First instar larvae (L1) are small and white and transform into yellow and slightly bigger second instar larvae (L2) after 1 or 2 days, depending on the temperature (Fig. 1). By synchronizing their development, a collection of larvae at a certain developmental stage (L1 or L2) can be obtained.

1. Fold some paper tissues into squares of 10 × 10 cm and place at the bottom of a 1-L container made in Subheading 3.2.1.
2. Rinse the bean pods with water and place five beans in each container.

3. Fill a PCR tube with bee pollen and close with its cap. Make a small hole in the bottom of the tube by cutting it with scissors or by piercing it with a needle, through which the thrips can acquire access to the pollen (*see* **Note 5**). Place one tube in the container.
4. Collect 100 female thrips as described in Subheading 3.3.1 and add them to the container (*see* **Note 6**). Allow them to oviposit on the beans for 48 h.
5. After 48 h, repeat **step 1** and place the oviposited beans in the female-free container. Add the bee pollen and put the females from **step 4** back into the BugDorm.
6. After 4–5 days 100–200 L1 larvae will emerge. They will develop into L2 larvae 2 or 3 days later.
7. Use a fine paintbrush to collect an L1 or L2 larva. Moisturizing the brush will improve sticking of the larva to the brush.
8. Gently place the hairs of the brush on a leaf or in the clip cage and allow the larva to move from the brush. Prevent brushing the larva off as this might damage the thrips.

3.4 Thrips Assays

3.4.1 Feeding Damage Assays

Measuring thrips feeding damage is a good measure to quantify thrips performance. Thrips silver damage can be estimated in mm^2 by eye or quantified from pictures using software [9]. Different thrips developmental stages can be used for this assay, but it should be kept in mind that the duration of the different stages differs (Fig. 1). When starting with L1 larvae, damage can be recorded for 5 days before the pupae phase starts. Damage caused by L2 larvae can therefore be recorded only for 3 days. The damage inflicted by female adults must be recorded within 4 days, before their eggs hatch and L1 larvae start to feed from the plant.

Whole-Plant Damage Assay

1. Test at least ten replicates per treatment and/or genotype. Put 10 mL of water in the petri dish at the bottom of a handmade container (*see* Subheading 3.2.1). Place an *A. thaliana* plant on the petri dish.
2. Collect thrips at the preferred stage as described in Subheading 3.3.
3. Treat the plant with the thrips (*see* **Notes** 7 and **8**).
4. Monitor the amount of feeding damage (see above; Fig. 6a).

Single-Leaf Damage Assay

1. Test at least ten replicates per treatment and/or genotype. Use clip cages (*see* Subheading 3.2.2).
2. Collect thrips at the preferred stage as described in Subheading 3.3.
3. Transfer the thrips to the clip cage (*see* **Note 9**).

Fig. 6 Silver damage on an *A. thaliana* plant (**a**) and application of a clip cage on an *A. thaliana* leaf (**b**)

4. Gently open the clip and place the leaflet in the opening of the clip cage. Gently release the grip on the clip so that the leaflet is fixed in the clip (Fig. 6b).
5. Let the clip rest on a neighboring pot to prevent damage on the petiole because of the weight of the clip.
6. Monitor the amount of feeding damage (see above; Fig. 6a; *see* **Note 10**).

3.4.2 Oviposition Assays

Feeding damage on the plant often correlates with oviposition by female thrips. As an extra measure of thrips performance, the amount of oviposited eggs can be determined after feeding by adult females. Thrips eggs are kidney shaped and are oviposited in leaf blades and petioles (Fig. 7).

1. After damage assessment of a whole-plant assay (*see* Subheading "Whole-Plant Damage Assay"), rosettes are harvested by cutting the plant at the hypocotyl. Fold the leaves upward like an umbrella and put the rosette in a 15-mL tube. When assaying single leaves (using clip cages), cut the treated leaf and place it in a 1.5-mL safe-lock tube.
2. Heat up a water bath to 100 °C.
3. Add the Trypan blue solution to the leaf tissue (~10 mL for a rosette, ~0.75 mL for a leaf) and close the tubes with their respective caps.

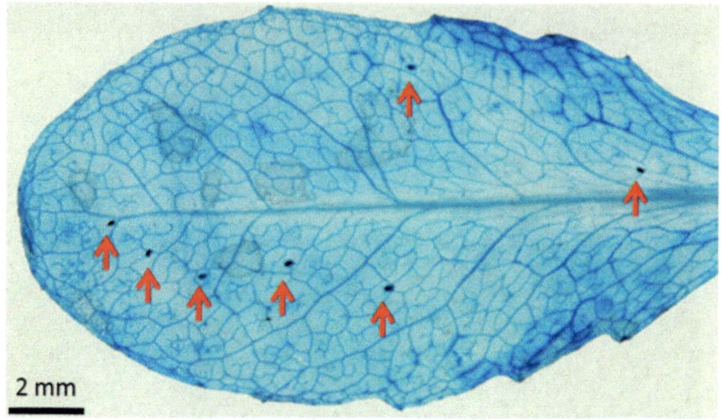

Fig. 7 Trypan blue-stained leaf to visualize thrips eggs. Red arrows point to oviposited eggs on an *A. thaliana* leaf

4. Place the 15-mL tubes for 2 min and the 1–5 mL tubes for 30 s in the water bath. Let the solution cool down for 30 min.
5. Pour off the Trypan Blue solution (use forceps to prevent the leaf material from sliding out of the tube) and replace by the lactoglycerol solution until the leaf tissue is fully submerged (~7 mL for a rosette, ~0.75 mL for a leaf). Samples can be stored at room temperature.
6. Place the rosette in a petri dish (Ø 15 cm) containing 95% ethanol. Single leaves are transferred to a microscope glass slide. Use a stereo zoom microscope and count the detected eggs. A magnification of 10× usually suffices, but a higher magnification may be needed if the level of background staining is high.

3.4.3 Assays for Thrips-Induced Changes in Gene Expression or Metabolite Accumulation

For measuring the plant's molecular responses to thrips, for example, induced changes of gene expression or metabolite accumulation, the use of a single leaf that remains attached to the plant until the point of harvest for the assay is preferred. The variation between replicates will be reduced by selecting leaves of a similar developmental stage. Moreover, by using a clip cage, a defined number of thrips can stay localized on one leaf (Fig. 3).

1. Follow **steps 1–5** of Subheading "Single-Leaf Damage Assay". Instead of ten replicates, this type of assays requires a minimum of five replicates.
2. Harvest the sample after the preferred period of time (*see* **Note 11**) by gently removing the clip and cutting the treated leaf at the petiole. Put the leaf in a 1.5-mL safe-lock tube and snap freeze it in liquid nitrogen, after which it can be stored in the −80 °C freezer.

4 Notes

1. It is important to have separate areas for plant cultivation, thrips rearing, and plant–thrips assays. Thrips are difficult to control, since they can easily maneuver through small openings and consequently contaminate other areas. In our case, we performed all our thrips assays in greenhouse compartments, which were physically separated from our growth chambers (Sylvania Luxline plus lamps F58W/84) where we usually cultivated our *A. thaliana* plants until they were moved to the greenhouse at least 24 h before the start of the thrips assay. Thrips rearing took place in a separate compartment in the greenhouse. In the beginning, we also cultivated *A. thaliana* plants in the greenhouse, but the frequency of contamination with thrips became too high.
2. The garden felt will retain the water longer, creating a moist environment for plants for a longer period.
3. No garden felt was used on the bottom of the trays in which lettuce plants were grown. After a few weeks, the roots of lettuce start to grow out from under the pots and attach to the felt. Removing the pots from the felt will damage the roots. This root outgrowth does not occur with *A. thaliana* plants.
4. Foam prevents the leaf from damaging. Additionally, foam closes openings between the two rings at the side where the petiole is placed between rings. Also, if the cages would be used on leaves of other plant species that might have thicker veins, the foam will close the spaces/openings that are created by the uneven thickness of the leaf.
5. Bee pollen are highly nutritious for thrips and can have a positive effect on, for example, development time, fecundity, and longevity.
6. A 15-mL pot of the aspirator containing the 100 female thrips can be opened and placed inside the 1-L rearing container with beans.
7. When a whole plant is treated with adult (female) thrips, the 15-mL pots of the aspirator can be opened and placed next to the petri dish inside the 1-L container.
8. Taken over several days, in our experiments, the average temperature inside the 1-L containers was 1.06 and 0.32 °C higher than outside the container during the day and night, respectively.
9. To get the female adults in a clip cage, tap against the 15-mL collection pot and hold it upside down above the open clip cage to release the thrips.

10. Gently remove the clip cage from the leaf and keep the leaf attached to the plant when assessing the damage. This will prevent wilting of the leaf, which hinders the scoring.
11. We harvested samples for analyses of gene expression and hormone accumulation every other hour until a maximum duration of two days (leaves will be too heavily infested after that time).

Acknowledgments

The authors would like to thank Hans van Pelt for the nice pictures. The authors are supported by the STW Perspective program Green defence Against Pests, which is partly financed by the Netherlands Organisation for Scientific Research (NWO) and the companies Bejo Zaden, East-West Seed, Enza Zaden, KeyGene, Rijk Zwaan, and Syngenta Seeds.

References

1. Steenbergen M, Abd-el-Haliem A, Bleeker P et al (2018) Thrips advisor: exploiting thrips-induced defences to combat pests on crops. J Exp Bot 69:1837–1848
2. Tommasini MG, Maini S (1995) *Frankliniella occidentalis* and other thrips harmful to vegetable and ornamental crops in Europe. Wageningen Agr Univ Pap 95:1–42
3. Reitz SR (2009) Biology and ecology of the western flower thrips (Thysanoptera: Thripidae): the making of a pest. Fla Entomol 92:7–13
4. Chisholm IF, Lewis T (1984) A new look at thrips (Thysanoptera) mouthparts, their action and effects of feeding on plant tissue. Bull Entomol Res 74:663–675
5. Kindt F, Joosten NN, Peters D et al (2003) Characterisation of the feeding behavior of the western flower thrips in terms of electrical penetration graph (EPG) waveforms. J Insect Physiol 49:183–191
6. Abe H, Ohnishi J, Narusaka M et al (2008) Function of jasmonate in response and tolerance of *Arabidopsis* to thrip feeding. Plant Cell Physiol 49:68–80
7. Bruessow F, Gouhier-Darimont C, Buchala A et al (2010) Insect eggs suppress plant defence against chewing herbivores. Plant J 62:876–885
8. Thoen MPM, Kloth KJ, Wiegers GL et al (2016) Automated video tracking of thrips behavior to assess host-plant resistance in multiple parallel two-choice setups. Plant Methods 12:1–12
9. Visschers IGS, Van Dam NM, Peters JL (2018) An objective high-throughput screening method for thrips damage quantification using Ilastik and ImageJ. Entomol Exp Appl 166:508–515
10. Maharijaya A, Vosman B, Steenhuis-Broers G et al (2011) Screening of pepper accessions for resistance against two thrips species (*Frankliniella occidentalis* and *Thrips parvispinus*). Euphytica 177:401–410
11. Hoagland DR, Arnon DI (1938) The water culture method for growing plants without soil. Calif Agric Exp Stn Bull 347:36–39.
12. Koch E, Slusarenko A (1990) Arabidopsis is susceptible to infection by a downy mildew fungus. Plant Cell 2:437–445

Chapter 8

Role of Jasmonate in Modulation of Mycorrhizae-Induced Resistance Against Fungal Pathogens

Aswathy Nair, H. V. Thulasiram, and Sujata Bhargava

Abstract

Symbiotic association of plants with arbuscular mycorrhizal (AM) fungi brings about changes in levels of the phytohormone jasmonate (JA) in root and shoot tissues of a plant. The enhanced JA levels not only play a role in controlling the extent of AM colonization but are also involved in the expression of mycorrhizal-induced resistance (MIR) against pathogens. We describe a method used to study the levels of a volatile jasmonate derivative, methyl jasmonate (MeJA), in tomato plants colonized by AM fungi and in response to subsequent attack by the foliar pathogen *Alternaria alternata*.

Key words Gas chromatography, Internal standard, Methyl jasmonate

1 Introduction

Arbuscular mycorrhizal (AM) symbiosis is widely distributed in the plant kingdom and has proven to be ecologically and agriculturally important in terrestrial ecosystems. The symbiosis involves exchange of mineral nutrients and sugar between the AM fungus and host plant, respectively. The initial response of host plant to the AM fungus is similar to that occurring during plant–pathogen interactions, and a number of redundant signaling pathways are induced that lead to the expression of defense responses. The defense responses constitute localized responses that reduce the onset of disease at sites of infection, as well as systemic responses that impart resistance to other parts of the plant against subsequent infection [1]. The phytohormones salicylic acid (SA) and jasmonate (JA) are known to play a role in the expression of defense responses through activation of transcription factors that regulate expression of defense-related genes [2].

During the early stages of establishment of symbiosis, the AM fungus is recognized as a biotrophic pathogen by the host plant, and endogenous SA levels increase, triggering the expression of plant defense genes [3]. The AM fungus suppresses the

Antony Champion and Laurent Laplaze (eds.), *Jasmonate in Plant Biology: Methods and Protocols*, Methods in Molecular Biology, vol. 2085, https://doi.org/10.1007/978-1-0716-0142-6_8, © Springer Science+Business Media, LLC, part of Springer Nature 2020

SA-mediated defense responses for establishing a symbiotic association. During later stages of symbiosis, JA levels increase, in both the roots [4] and shoots [5] of AM-colonized plants. JA is thought to play a role in limiting the growth of the AM fungus, such that its association with the plant remains mutualistic and does not become pathogenic [6, 7].

AM fungal association has also been shown to evoke stronger and faster defense responses in the host plant in response to subsequent pathogen attack [8]. This priming effect of the AM fungal association leads to the expression of mycorrhizal-induced resistance (MIR; [9, 10]). MIR is widely reported against soil-borne pathogens and in some cases against above-ground or shoot pathogens [11]. Although the mechanism of MIR has not been understood fully, it is thought to be mediated through various biochemical and molecular alterations observed in AM-colonized plants. The high JA levels observed in AM-colonized plants may contribute to the expression of MIR. In AM-colonized tomato plants, the increase in JA levels arising from the symbiotic association were possibly involved in the expression of MIR against the necrotrophic pathogen *Alternaria alternata* and the biotrophic pathogen *Fusarium oxysporum* pv. *lycopersici* [12, 13]. Increased expression of genes involved in JA biosynthesis and response was also observed in the AM-colonized plants subjected to pathogen application. In this chapter, we will discuss the experimental procedures involved in the establishment of AM symbiosis in tomato plants and the methodology used to estimate methyl jasmonate (MeJA, *see* **Note 1**) levels in response to AM colonization and in the expression of MIR against *A. alternata*.

2 Materials

2.1 Plant Material, AM Fungal Inoculum and Pathogen

1. Seeds of tomato, *Solanum lycopersicum* L. (cv. Pusa Ruby).
2. Soil-based inoculum of *Glomus fasiculatum*.
3. *Alternaria alternata* pv. *lycopersici* growing on Potato Dextrose Agar medium (available commercially).
4. Compound microscope.

2.2 Staining of Mycorrhizal Colonized Roots

1. 10% KOH.
2. 1% HCl.
3. Acidified glycerol solution (50 ml glycerol, 45 ml water, 5 ml 1% HCl).
4. 0.05% Trypan blue stain prepared in acidified glycerol.

2.3 Extraction and Quantitation of Methyl Jasmonate

1. HPLC grade (95%), Methyl jasmonate.
2. Analytical reagent grade: acetone, methanol, and dichloromethane.
3. Dichloromethane acidified to pH 2.0 with 6 M HCl.
4. Mortar and Pestle.
5. Sonicator.
6. Centrifuge.
7. Speed vacuum concentrator.
8. Gas chromatograph.
9. HP-5 column (30 m × 0.32 mm × 0.25 mm).
10. Nitrogen as a carrier gas for chromatography.
11. Glass syringe for injecting 5 μl of sample.

3 Methods

3.1 Plant Material, AM Fungal Inoculation and Pathogen Application

1. Raise tomato, *S. lycopersicum* L. (cv. *Pusa Ruby*) seedlings in seedling trays and transplant after 20 days to bags containing about 1 kg of autoclaved soil or potting mixture.
2. Apply about 5 g by volume of the soil-based *G. fasiculatum* inoculum (containing >1 spore g^{-1}) in the small pit made for transplantation of tomato seedlings to the soil filled bags (Treatment—AM). Plants to which no mycorrhizal inoculum is applied are Controls (C).
3. Grow the plants in a greenhouse under diffused daylight and temperature between 25 and 28 °C. After 45 days, confirm AM colonization from a sample of fine lateral roots by clearing the roots in 10% potassium hydroxide (KOH), acidifying them with 1% hydrochloric acid (HCl) and then staining them with 0.05% Trypan blue (http://invam.caf.wvu.edu/methods/mycorrhizae/staining.htm).
4. At 45 days after sowing, transfer the tomato plants to a small polythene tent within the same greenhouse to increase humidity. Apply 200 μl spore suspension (approximately 5×10^3 spores) prepared from a culture of *A. alternata* to the adaxial surface of the first fully expanded leaf of control (C) and AM-inoculated (AM) plants using a micropipette. The pathogen treatments would be C + P and AM + P, respectively.
5. After 3 days of spore application, collect tissue samples for estimating MeJA levels. Score the percent disease index (PDI), 5–7 days after spore application (*see* **Notes 2–4**).

3.2 Extraction of Methyl Jasmonate

1. Weigh fresh tissue (1 g) and grind to a fine powder in liquid nitrogen.

2. Add 1 ml of 80 % methanol to the ground powder in a microcentrifuge tube and incubate overnight at 4 °C.
3. Sonicate the samples for 20 min, centrifuge at 8000 × *g* for 10 min and re-extract the pellet in 1 ml of acetone:methanol (1:1). Pool the two supernatant fractions and evaporate the organic solvents in a speed vacuum concentrator leaving the aqueous phase in the microcentrifuge tube.
4. Partition the aqueous phase twice using acidified dichloromethane.
5. Take the dichloromethane layer and evaporate the solvent at room temperature, using a speed vacuum concentrator. Dissolve the residue in 50 μl of methanol.
6. Use this methanolic extract for quantitation of MeJA.

3.3 Quantitation of Methyl Jasmonate

1. Estimate MeJA levels in the plant tissue using gas chromatography (Agilent 7890, Singapore) with an HP-5 column (30 m × 0.32 mm × 0.25 mm, J and W Scientific).
2. Use nitrogen as the carrier gas at a flow rate of 1 ml min^{-1}.
3. Inject 5 μl of the methanolic extract through the injector port (*see* **Notes 5** and **6**).
4. Maintain the column at 80 °C for 2 min. Then, program a gradient increase in temperature from 80 °C to 150 °C at the rate of 2 °C min^{-1}, followed by a further temperature increase to 220 °C at the rate of 15 °C min^{-1}, and finally maintain the column temperature at 220 °C for 10 min. The temperature of the injector port and detector is maintained at 250 °C.
5. Use a flame ionization detector (FID) for scoring the peaks.
6. Confirm the peak corresponding to MeJA by mass spectrometry (Fig. 1).
7. Prepare a calibration curve, using peak areas versus different concentrations of the hydrogenated MeJA, which is used as an internal standard (*see* **Notes** 7 and **8**) (Fig. 2).
8. Add 80 ng of internal standard to plant tissue samples at the first step of MeJA extraction.
9. Use the peak area of internal standard added to plant tissue samples to quantitate MeJA levels in the samples and also to determine percent recovery of MeJA (*see* **Note 9**).
10. Substitute the peak area of MeJA for (y) in the regression equation obtained for the calibration curve of internal standard (Fig. 2) and determine the value of (x), which is the concentration of MeJA in ng ml^{-1} of extract.
11. Multiply the MeJA concentration obtained with percent recovery of internal standard to determine actual concentration of MeJA in the extract.

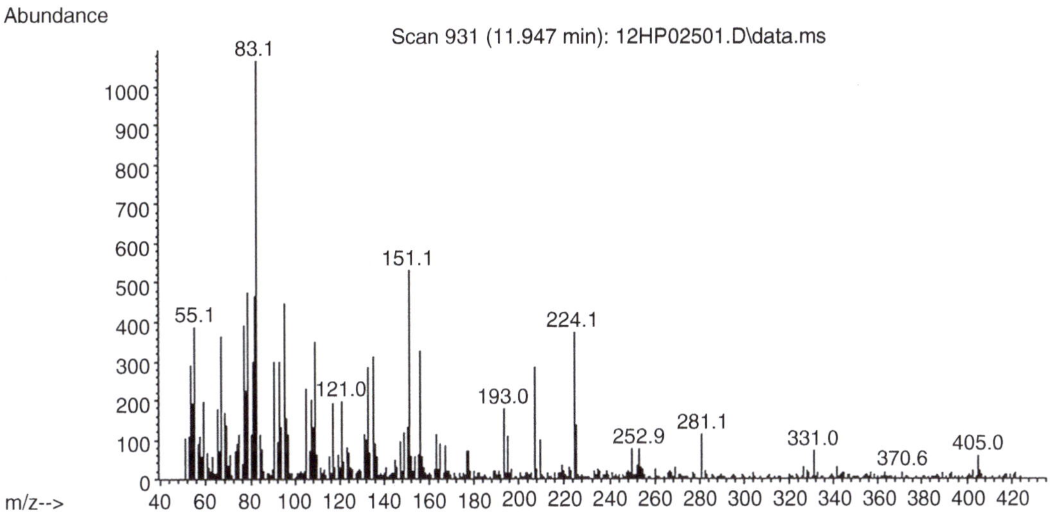

Fig. 1 Mass spectrometry spectrum of MeJA

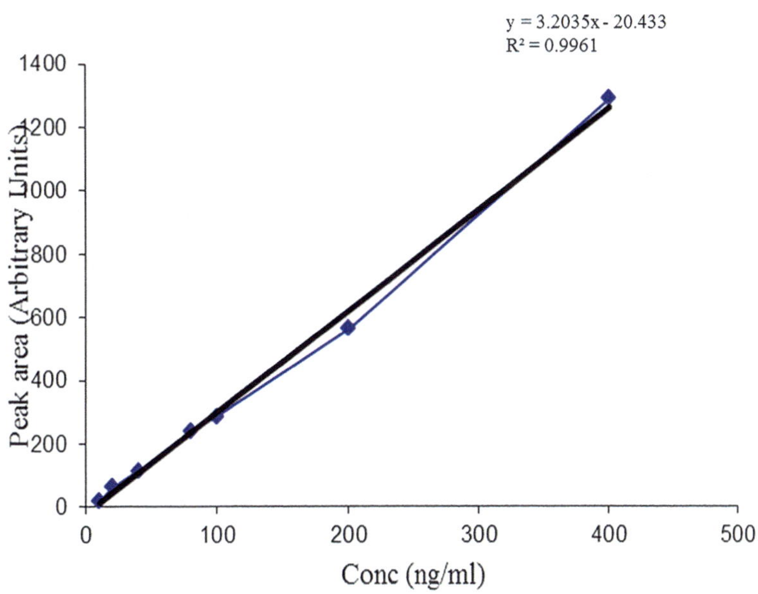

Fig. 2 Calibration curve (peak area vs. concentration of internal standard (hydrogenated MeJA)

12. Express MeJA concentration as ng MeJA g^{-1} tissue. Compare the MeJA levels in (C) and (AM) plants to see how AM colonization affects MeJA levels.
13. Correlate the MeJA levels to percent disease index (PDI) scored in control (C + P) and AM-colonized (AM + P) plants as an estimate of the expression of MIR.

4 Notes

1. MeJA and JA are interconvertible forms collectively called jasmonates. MeJA is formed by the action of a methyl transferase (JMT) on JA and is a volatile derivative of JA. Both forms are known to trigger defense responses in plants.
2. The percentage disease index (PDI) is calculated by observing the lesions formed on the pathogen-inoculated leaf. The leaves are rated on a scale of 0–5 according to the severity of symptoms observed.
3. The details of scales are as shown below:

   0: No disease symptoms.

   1: A few spots toward tip covering 10% leaf area.

   2: Several dark-brown patches covering up to 20% leaf area.

   3: Several patches with paler outer zone covering up to 40% leaf area.

   4: Leaf blight covering up to 75% leaf area or breaking of the leaves from the center.

   5: Complete drying of the leaves or breaking of the leaves from the center.
4. PDI is calculated as [14]

$$\text{Percent disease index (PDI)} = \frac{\text{Sum of individual ratings}}{\text{Number of leaves} \times \text{Maximum examined disease scale}} \times 100$$

5. Wash the column by injecting methanol in place of the sample containing MeJA in a blank run. Then inject the next MeJA sample.
6. Take the same time to inject the samples so as to avoid variation in the retention time for the sample in different runs.
7. The internal standard, hydrogenated MeJA, is synthesized chemically. A total of 0.1 mg of MeJA in 5–10 ml HPLC grade methanol is stirred with 10 mg 10% Palladium on activated carbon under 2 Atm pressure of hydrogen. After 8–10 h, the reaction mixture is filtered through a celite bed and chromatographed on silica gel (230–400 mesh) column using a gradient mixture of methanol/dichloromethane as eluent to get the hydrogenated derivative of MeJA. Hydrogenated MeJA is confirmed using NMR. This derivative of MeJA is not present in plant tissues and gives a clearly separated peak from the peak corresponding to MeJA.
8. The calibration curve should cover the range of MeJA concentrations in the plant tissue samples.

9. The peak area of internal standard detected in the tissue extract is divided by the peak area (244 Arbitrary Units) corresponding to the actual concentration of internal standard added to the tissue sample (80 ng ml^{-1}). This value is expressed as a percentage and represents the percentage of internal standard recovered after the extraction procedure.

Acknowledgment

This work was carried out under a research grant provided by the Science and Engineering Research Board, Government of India.

References

1. Mandadi KK, Scholthof KB (2013) Plant immune responses against viruses: how does a virus cause disease? The Plant Cell 25:1489–1505
2. Alazem M, Lin NS (2015) Roles of plant hormones in the regulation of host-virus interactions. Mol Plant Pathol 16:529–540
3. Paszkowski U (2006) A journey through signalling in arbuscular mycorrhizal symbioses. New Phytol 172:35–46
4. Hause B, Mrosk C, Isayenkov S, Strack D (2007) Jasmonates in arbuscular mycorrhizal interactions. Phytochemistry 8:101–110
5. Kapoor R (2008) Induced resistance in mycorrhizal tomato is correlated to concentration of jasmonic acid. OnLine J Biol Sci 8:49
6. Hanlon MT, Coenen C (2011) Genetic evidence for auxin involvement in arbuscular mycorrhiza initiation. New Phytol 189:701–709
7. Gutjahr C, Paszkowski U (2009) Weights in the balance: jasmonic acid and salicylic acid signaling in root-biotroph interactions. Mol Plant-Microbe Interact 22:763–772
8. Jung SC, Medina AM, Lopez-Raez JA, Pozo MJ (2012) Mycorrhiza-induced resistance and priming of plant defenses. J Chem Ecol 38:651–664
9. Pozo MJ, Azcon-Aguilar C (2007) Unraveling mycorrhiza-induced resistance. Curr Opin Plant Biol 10:393–398
10. Pozo MJ, Jung SC, Lopez-Raez JA, Azcon-Aguilar C (2010) Impact of arbuscular mycorrhizal symbiosis on plant response to biotic stress: the role of plant defence mechanisms. In: Koltai H, Kapulnik Y (eds) Arbuscular mycorrhizas: physiology and function. Springer, Heidelberg, pp 193–207
11. Whipps JM (2004) Prospects and limitations for mycorrhizas in biocontrol of root pathogens. Can J Bot 82:1198–1227
12. Nair A, Kolet SP, Thulasiram HV, Bhargava S (2015a) Systemic jasmonic acid modulation in mycorrhizal tomato plants and its role in induced resistance against *Alternaria alternata*. Plant Biol 17:625–631
13. Nair A, Kolet SP, Thulasiram HV et al (2015b) Systemic jasmonic acid modulation in mycorrhizal tomato plants and its role in induced resistance against Alternaria alternata. Plant Biol 17:625–631
14. Wheeler BEJ (1969) An Introduction to plant diseases. John Wiley and Sons Ltd., London, p 254

Chapter 9

Establishment of Actinorhizal Symbiosis in Response to Ethylene, Salicylic Acid, and Jasmonate

Mariama Ngom, Maimouna Cissoko, Krystelle Gray, Valérie Hocher, Laurent Laplaze, Mame Ourèye Sy, Sergio Svistoonoff, and Antony Champion

Abstract

Phytohormones play a crucial role in regulating plant developmental processes. Among them, ethylene and jasmonate are known to be involved in plant defense responses to a wide range of biotic stresses as their levels increase with pathogen infection. In addition, these two phytohormones have been shown to inhibit plant nodulation in legumes. Here, exogenous salicylic acid (SA), jasmonate acid (JA), and ethephon (ET) were applied to the root system of *Casuarina glauca* plants before *Frankia* inoculation, in order to analyze their effects on the establishment of actinorhizal symbiosis. This protocol further describes how to identify putative ortholog genes involved in ethylene and jasmonate biosynthesis and/or signaling pathways in plant, using the *Arabidopsis* Information Resource (TAIR), Legume Information System (LIS), and Genevestigator databases. The expression of these genes in response to the bacterium *Frankia* was analyzed using the gene atlas for *Casuarina–Frankia* symbiosis (SESAM web site).

Key words Ethylene, Jasmonate, Actinorhizal symbiosis, *Frankia*, *Casuarina glauca*

1 Introduction

Plant hormones are a group of naturally occurring, organic molecules, which influence physiological processes such as plant growth, differentiation, and development at low concentrations [1]. Among them, salicylic acid, jasmonate, and ethylene are mainly known to play crucial roles in regulating plant defense responses to a wide range of biotic and abiotic stresses [2–4]. Salicylic acid is one of numerous plant phenolic compounds that are a derivative of cinnamic acids [5, 6]. It is generally involved in plant defense–related processes against biotrophic and hemi-biotrophic pathogens, as well as the establishment of systemic acquired resistance (SAR) [2]. Increase in salicylic acid levels in pathogen-exposed tissues results in the induction of pathogenesis-related (PR) genes and enhanced resistance to a broad range of pathogens [4]. By

Antony Champion and Laurent Laplaze (eds.), *Jasmonate in Plant Biology: Methods and Protocols*, Methods in Molecular Biology, vol. 2085, https://doi.org/10.1007/978-1-0716-0142-6_9,

contrast to salicylic acid, jasmonate and ethylene are often associated with defense reactions against necrotrophic pathogens and herbivorous insects [7, 8]. Jasmonate and its bioactive form jasmonate Isoleucine (JA-Ile) are synthetized from polyunsaturated fatty acids like linoleic acid in *Arabidopsis thaliana* [9]. Several studies have demonstrated that endogenous concentrations of jasmonate increase locally in response to pathogen infection or tissue damage and its exogenous application induced the expression of defense-related genes [10–13]. The molecular components of jasmonate biosynthesis and signaling pathway have been extensively investigated in both dicot- and monocotyledonous plants, and most of genes involved in these processes were identified including OPR3, JAZ1, JAZ3, and ORA47 [14–17]. Ethylene is the simplest olefin that is produced either chemically through the incomplete combustion of hydrocarbons or biologically by almost all living organisms [1]. In plants, this gaseous hormone is synthesized from the methionine and 1-aminocyclopropane-1-carboxylic acid (ACC) is its metabolic precursor [18]. Similar to the defense-regulating compounds—salicylic acid and jasmonate—plant-derived ethylene is also considered to be involved in resistance to pathogens [19]. Nevertheless, it is well established that the precise activation of a set of plant defense responses is not induced only by one of these phytohormones, but it is the result of a network of interactions between different signaling pathways, especially those of ethylene and jasmonate [2, 4, 19–21].

Because phytohormones are signal molecules involved in plant–microorganism interactions, they are likely to positively or negatively regulate the establishment of nitrogen-fixing root–nodule symbiosis. In legumes, ethylene and jasmonate have been shown to inhibit plant nodulation, calcium spiking in root hairs, and the expression of two early symbiotic genes *RIP1* and *ENOD11* [22–24]. Salicylic acid inhibited indeterminate-type nodulation but not determinate-type one [25]. To date, little is known regarding the involvement of plant stress hormones on nodule development in *Casuarina glauca*, which is also able to form a nitrogen-fixing symbiosis with actinobacteria of the genus *Frankia*. *C. glauca* is a model tree for basic and fundamental research in actinorhizal symbiosis with the development of many tools, including transcriptome analyses and a gene atlas for *Casuarina–Frankia* symbiosis [26–29].

Here, exogenous salicylic acid (SA), jasmonate acid (JA), and ethephon (ET) were applied to plant roots before *Frankia* inoculation in order to analyze their effects on both *C. glauca* nodulation and root hair deformation as one of the early steps of the infection process, leading to the development of root nodules. Using the *Arabidopsis* Information Resource (TAIR), Legume Information System (LIS), and Genevestigator databases, putative ortholog genes involved in ethylene and jasmonate biosynthesis and/or signaling pathways in *C. glauca* were identified and their expression in response to the bacterium *Frankia* were then analyzed, thanks to the gene atlas for *Casuarina–Frankia* symbiosis (SESAM web site).

2 Materials

2.1 Plant Material

1. *Casuarina glauca* seeds (lot 15,934, ref.086-5929).

2.2 Bacteria Material

1. *Frankia casuarinae* CcI3 [30].

2.3 Chemicals, Solutions, Media and Equipment

1. Phytohormones: Jasmonic acid, ethephon, and salicylic acid.
2. Media: Buffer Added Phosphate medium [31], supplemented with 0.5 mM-pH 6.8 MES–Tris buffer, 1 M sodium propionate and 1 M NH_4Cl (*see* **Note 1**); Broughton and Dilworth medium [32] and Hoagland medium [33].
3. Bacterial culture facilities.
4. Media flasks (1 L) and glass tubes (50 mL).
5. Sterile polypropylene tubes (50 mL), plastic serological pipettes (10 mL), Terumo needle (0.8 × 40 mm), and syringe (10 mL).
6. 0.20 μm sterile hydrophilic filters.
7. Sterile Petri dishes (245 × 245 mm).
8. Sterile ddH_2O.
9. Growth chamber with appropriate lighting and temperature control (conditions: 28 ± 1 °C, 16-h day/8-h night photoperiod, 74 μmol m^{-2} s^{-1} light intensity).
10. BX50F microscope with a Micro Publisher 3.3 RTV digital camera.

2.4 Bacterial Cultures and Inoculum Preparation

1. Bacteria were grown into flasks containing 200 mL of the sterile liquid BAP medium at a final pH of 6.7. Cultures were slowly stirred and maintained in darkness at 28 ± 1 °C. After 14 days of cultivation, the supernatant was removed and bacterial cells were collected by centrifugation at 2000 × *g* for 15 min. Pellets were washed twice with 40 mL of sterile ddH_2O and homogenized by repeated aspirations and discharges using sterile needle and syringe.
2. For inoculum preparation, bacteria homogenized cells were suspended in sterile ddH_2O with a final absorbance of 0.2 (*see* **Note 2**), measured at $\lambda = 595$ nm.

2.5 Plant Cultures

1. *C. glauca* seeds were germinated under semi-axenic conditions in a plastic tray (53.5 × 27.5 cm), containing a sterile mixture of compost and sandy soil (v/v;120 °C, 60 min). They were

watered daily with a quarter-strength Hoagland liquid medium. One month after seed germination, seedlings were uprooted from the soil, gently washed five times with ddH_2O. They were individually transferred in hydroponic conditions, into glass tubes filled with a 50-mL liquid BD medium supplemented with 5 mM KNO_3 as nitrogen source, at pH 6.7. Seedlings were incubated in a growth chamber. The liquid BD medium was renewed every 2 weeks (*see* **Note 3**).

2. After 1 month in hydroponic conditions, the plants were placed in nitrogen-free-BD medium.

2.6 Databases and Software for the Identification of Ortholog Genes and Expression Analysis

1. *Arabidopsis* Information Resource (TAIR, https://www.arabidopsis.org) [34].
2. Genevestigator database [35].
3. Legume Information System (LIS, https://legumeinfo.org) [36].
4. Gene atlas for *Casuarina–Frankia* symbiosis (https://www.genoscope.cns.fr/agc/sesam-dev/home/index.php).
5. BioEdit, CLC, SeaView and FigTree softwares.

3 Methods

3.1 Plant Roots Elicitation with SA, JA and ET

1. Prepare nitrogen-free BD medium containing the following final concentrations: 50, 100, 200, and 400 μM SA; 5, 50, and 500 μM ethephon (ET); 1, 50, and 100 μM of jasmonic acid (JA); and 500 μM ET added to 50 μM JA (*see* **Note 4**).
2. Remove the BD medium from the 5-day-old nitrogen-deficient plants and replace it with the nitrogen- free BD medium containing the required concentrations of phytohormones. A set of non-treated plants serves as control.
3. Incubate plants for 48 h in a growth chamber with appropriate lighting and temperature control.

3.2 Plant Nodulation Analysis

1. After 2 days of elicitation, completely remove the nitrogen-free BD medium, containing SA, ET, JA, and ET + JA.
2. Pour the bacterial inoculum previously prepared on sterile Petri dishes and place plant root system into contact with it for 2 h (*see* **Note 5**).
3. Replace back plants into tubes replenished with a nitrogen-free BD medium without phytohormones supplemented with 5 mL of the bacterial suspension.
4. Incubate plants in a growth chamber and follow their nodulation each week for about 7 weeks after inoculation.

5. Include at least two independent biological replicates with several plants for each treatment ($n = 24$) in the experimental design.

3.3 Plant Root Hair Deformation Analysis

1. After 2 days of elicitation, completely remove the nitrogen-free BD media, containing a final concentration of 500 μM ET, 50 μM JA, and 500 μM ET added to 50 μM JA.
2. Repeat **steps 2** and **3** described earlier.
3. Incubate plants in a growth chamber for 48 h.
4. Place plant root system delicately in Petri dishes containing sterile ddH_2O.
5. For each plant, take micrographs of three short lateral roots (*see* **Note 6**), using a BX50F microscope (10×) with Micro Publisher 3.3 RTV digital camera.
6. For each micrograph, evaluate root hair deformation intensity using the following scoring [37]: 0—no deformation; 1—straight root hair with tip swelling; 2—only one change in growth direction; 3—more than one change in growth direction but no bifurcation; 4—one or more bifurcations.
7. Include in the experimental design, at least two independent biological replicates with several plants for each treatment ($n = 5$).

3.4 Identification of Ortholog Genes in C. glauca

1. Identify from TAIR database, genes potentially involved in ethylene and jasmonate biosynthesis and/or signaling pathways in *Arabidopsis thaliana*:
   (a) Go to https://www.arabidopsis.org and use "ethylene" or "jasmonate" as search terms, select gene, and then press the "search" button (*see* **Note 7**).
   (b) Note the locus names of the genes found.
   (c) Download their FASTA full-protein sequences.
2. Select only genes that are regulated by ethylene or jasmonate treatments from the list of potential candidates, using Genvestigator engine.
3. Analyze *A. thaliana* genes found on LIS database:
   (a) Go to https://legumeinfo.org/blast/protein/protein and choose Blast Sequence Search > Blastp.
   (b) Copy and paste individually FASTA full-protein sequences of *A. thaliana* genes in the section "enter FASTA sequence".
   (c) Select "All proteins" in the section "Protein Blast databases", then press the "Blast" button.
   (d) Download in FASTA format the results of the alignment.

4. Align protein sequences of *A. thaliana* genes with the *C. glauca* proteome to identify potential ortholog genes:
   (a) Open BioEdit software, choose Accessory Application > Blast > Local Blast.
   (b) Select "Blastp" in the program and "*C. glauca*" proteome as protein database, then copy and paste individually FASTA full-protein sequences of *A. thaliana* genes in the query section.
   (c) Leave the default settings and press "Do search" button.
   (d) Note the results of the search (*e*-value, score, identities, positives, gaps, sequence length, and gene ID of potential ortholog genes found in *C. glauca)* and download FASTA protein sequences of the genes found from the *C. glauca* proteome using CLC software.
5. Build phylogenetic trees using LIS-uploaded files to which the individual protein sequences of the potential ortholog genes identified in *C. glauca* have been added.
   (a) Open Seaview software, select Align > Alignment options > muscle.
   (b) Select sites > create site > Gblocks > ok > choose the first three options.
   (c) Choose Trees > PhyML > run.
6. Use Figtree software to edit obtained phylogenetic trees (data not shown). Identified putative ortholog genes involved in ethylene and jasmonate biosynthesis and/or signaling pathways in *C. glauca* are presented in Table 1.

3.5 Expression of Identified Ortholog Genes

1. Go to the *SESAM* website (https://www.genoscope.cns.fr/agc/sesam-dev/home/index.php).
2. Select "Blast" and then copy and paste individually the FASTA full-protein sequences of *C. glauca* putative ortholog genes in the section "Query for Blast."
3. Choose "tblastn" in the Blast program and "*Casuarina glauca*" as reference, then press the bouton "run."
4. Click on the contig sequence numbers that have about 100% of identities.
5. Press the *Display* button in front of both "Expression in control conditions and Expression during nodulation kinetic."

3.6 Observations

1. The application of exogenous concentrations of ET and JA on plant root system causes significant inhibition of *C. glauca* nodulation unlike SA (Fig. 1; Table 2); with additive effects of ET and JA (Fig. 2).

Table 1
Putative ortholog genes potentially involved in ethylene and jasmonate biosynthesis and/or signaling pathways identified in C. *glauca*

	Arabidopsis thaliana			*Casuarina glauca*							
	Gene name	Gene ID	Length	Gene name	Gene ID	Length	Expect	Score	Identities	Positives	Gaps
Ethylene	ETR2	AT3G23150	773	Cg_ETR2	Casgl116S07365	768	0.0	895 bits	477/792 (60%)	594/792 (75%)	43/792 (5%)
	ERS1	AT2G40940	613	Cg_ERS1	Casgl99S06274	636	0.0	899 bits	444/594 (74%)	520/594 (87%)	–
	ATMKK9	AT1G73500	310	Cg_ATMKK9	Casgl66S25836	317	e-116	416 bits	208/316 (65%)	249/316 (78%)	8/316 (2%)
	EBF2	AT5G25350	623	Cg_EBF2	Casgl23S11863	649	0.0	728 bits	372/651 (57%)	480/651 (73%)	30/651 (4%)
	EDF1	AT1G25560	361	Cg_EDF1	Casgl64S02381	367	e-118	422 bits	237/385 (61%)	282/385 (73%)	42/385 (10%)
Jasmonate	JAZ3	At3G17860	352	Cg_JAZ3	Casgl66S20850	370	5.00E-54	210 bits	159/382 (41%)	204/382 (53%)	66/382 (17%)
	JAZ1	At1g19180	253	Cg_JAZ1	Casgl8S22205	274	4.00E-54	209 bits	134/276 (48%)	169/276 (61%)	29/276 (10%)
	ORA47	At1g74930	195	Cg_ORA47	Casgl8S02712	198	1.00E-38	157 bits	92/187 (49%)	113/187 (60%)	14/187 (7%)
	OPR3	At2g06050	391	Cg_OPR3	Casgl8S22226	399	e-173	607 bits	306/396 (77%)	338/396 (85%)	10/396 (2%)
	ACX1	At4g16760	651	Cg_ACX1	Casgl227S11635	665	0.0	1090 bits	527/665 (79%)	593/665 (89%)	14/665 (2%)
	MFP2	At3g06860	725	Cg_MFP2	Casgl15S09060	725	0.0	1165 bits	577/725 (79%)	648/725 (89%)	

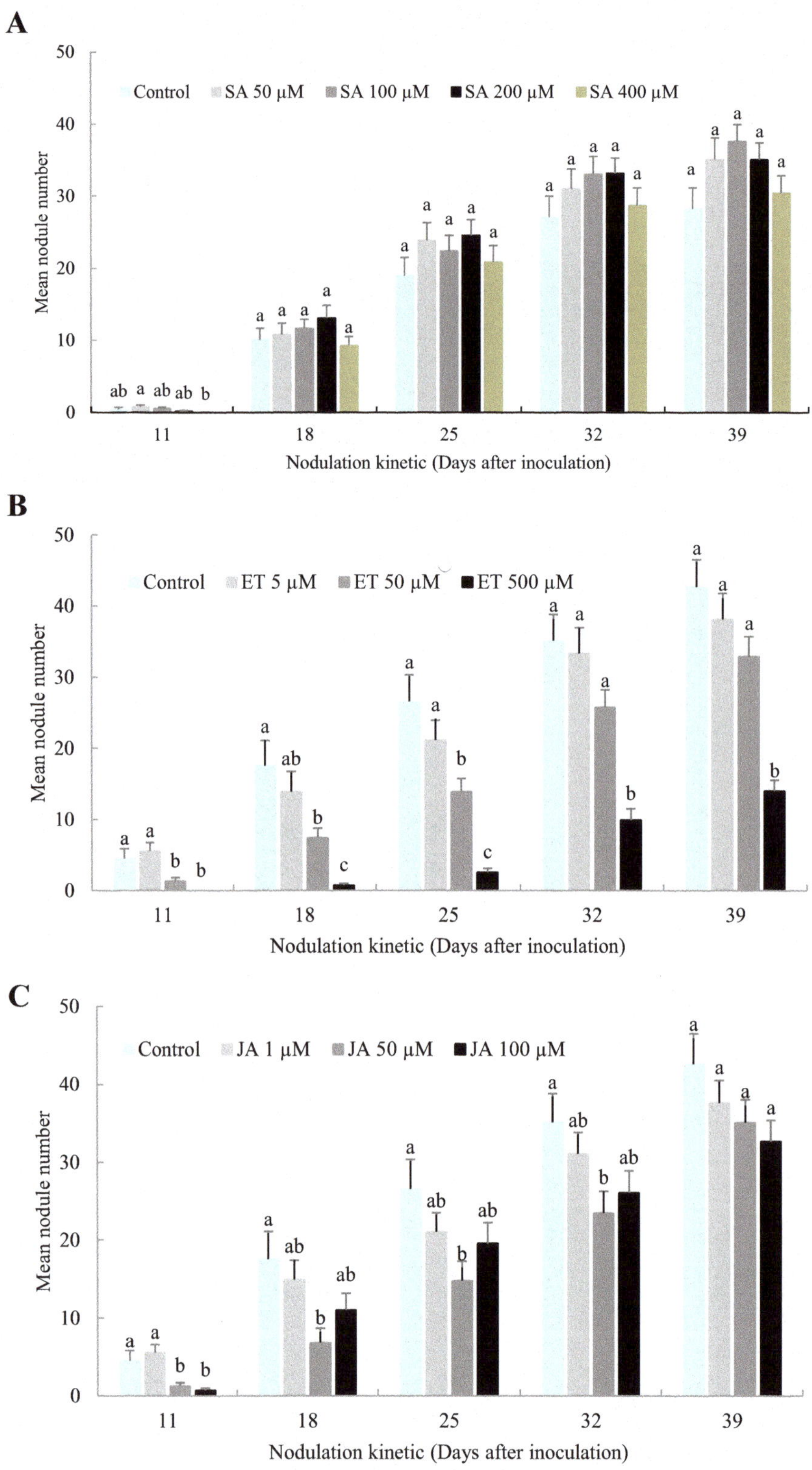

Fig. 1 Dose-rate effects of salicylic acid (**a**), ethephon (**b**) and jasmonic acid (**c**) on the nodulation of *C. glauca* plants by *F. casuarinae* Ccl3. Each value represents the mean number of plants used in each treatment ($n = 24$). Vertical bars indicate the standard error of mean. For each step of the nodulation kinetic, different lowercase letters (**a–c**) indicate significant difference between control and plants pretreated with salycilic acid (SA), ethephon (ET), or jasmonic acid (JA), according to the Student–Newman–Keuls (SNK) test at $P < 0.05$

Table 2
Dose-rate effects of salicylic acid (SA), ethephon (ET) and jasmonic acid (SA) on the nodulation rate in percentage of *C. glauca* plants nodulated by *Frankia casuarinae* Ccl3

	Nodulation rate (%) Days after inoculation				
Hormonal treatments	**11**	**18**	**25**	**32**	**39**
Control	12.5	95.8	95.8	100	100
SA 50 μM	37.5	95.8	95.8	100	100
SA 100 μM	33.3	95.8	100	100	100
SA 200 μM	16.7	95.8	100	100	100
SA 400 μM	0	95.8	100	100	100
Control	66.7	95.8	95.8	100	100
ET 5 μM	75	91.7	100	100	100
ET 50 μM	37.5	87.5	95.8	100	100
ET 500 μM	0	37.5	75	87.5	95.8
Control	66.7	95.8	95.8	100	100
JA 1 μM	75	100	100	100	100
JA 50 μM	41.7	66.7	95.8	100	100
JA 100 μM	29.2	66.7	87.5	95.8	100

Nodulation rate corresponds to the percentage of plant nodulated ($n = 24$)

2. The decrease in plant nodulation could be due to an inhibition of nodule initiation and/or infection processes, as ET and JA severely affect root hair deformation response to *Frankia* inoculation (Fig. 3).
3. In silico approaches used here allow for quick and reliable analyses that lead to the identification of 11 putative ortholog genes potentially involved in ethylene and jasmonate biosynthesis and/or signaling pathways in *C. glauca*. The analysis of their expression under symbiotic conditions reveals that several genes are differentially regulated, especially downregulated during the early stages (24–48 h after inoculation), whereas most of them are upregulated during the later stages of nodule formation (14 days after inoculation and in nodule) (Fig. 4).

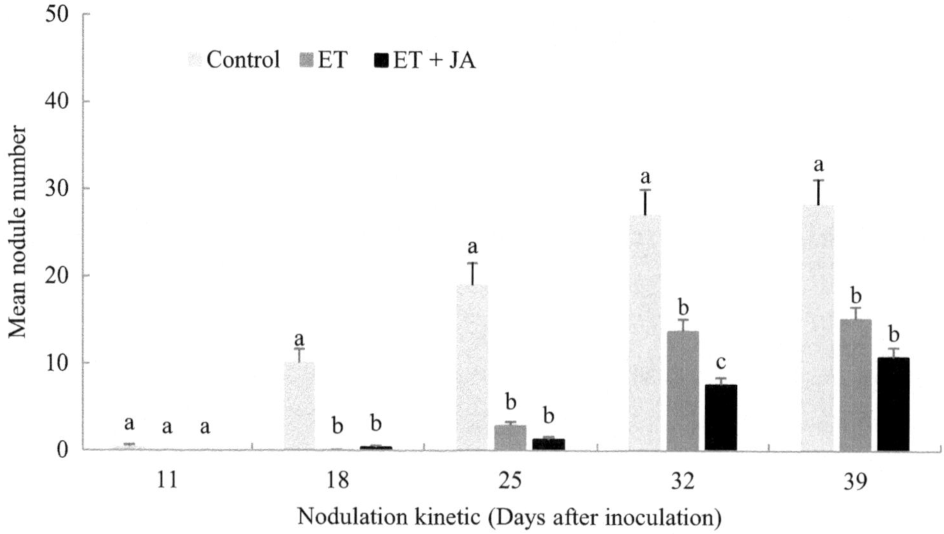

Fig. 2 Additive effects of ethephon and jasmonic acid on the nodulation of *C. glauca* plants by *F. casuarinae* CcI3. Each value represents the mean number of plants used in each treatment ($n = 24$). Vertical bars indicate the standard error of mean. For each step of the nodulation kinetic, different lowercase letters (**a–c**) indicate significant difference between control and plants pretreated with ethephon (ET) or ethephon added to jasmonic acid (ET + JA), according to the Student–Newman–Keuls (SNK) test at $P < 0.05$

4 Notes

1. Sodium propionate and NH_4Cl were added to the medium as carbon and nitrogen sources, respectively, for better growth of *Frankia* strain CcI3.
2. A 0.2 optical density of *Frankia* cells had previously allowed a good nodulation [38, 39].
3. The liquid BD medium for plant culture should be renewed every 2 weeks to avoid nutrient depletion and pH drift.
4. For elicitation studies with the aim of analyzing plant nodulation, non-physiologically high concentrations of stress phytohormones are often used [23].
5. When nodulating actinorhizal plants, we usually place plant roots into contact with *Frankia* cell suspensions for approximately 2 h to allow the attachment of bacteria on root hairs and thus the establishment of symbiosis as *Frankia* is a non-motile bacterium.
6. It is recommended to take micrographs from the apex of short lateral roots because an extensive deformation of root hairs is generally observed in this area.
7. It is necessary to associate to research on TAIR database a bibliographical research to consider only genes that are actually involved in ethylene or jasmonate biosynthesis and signaling pathways.

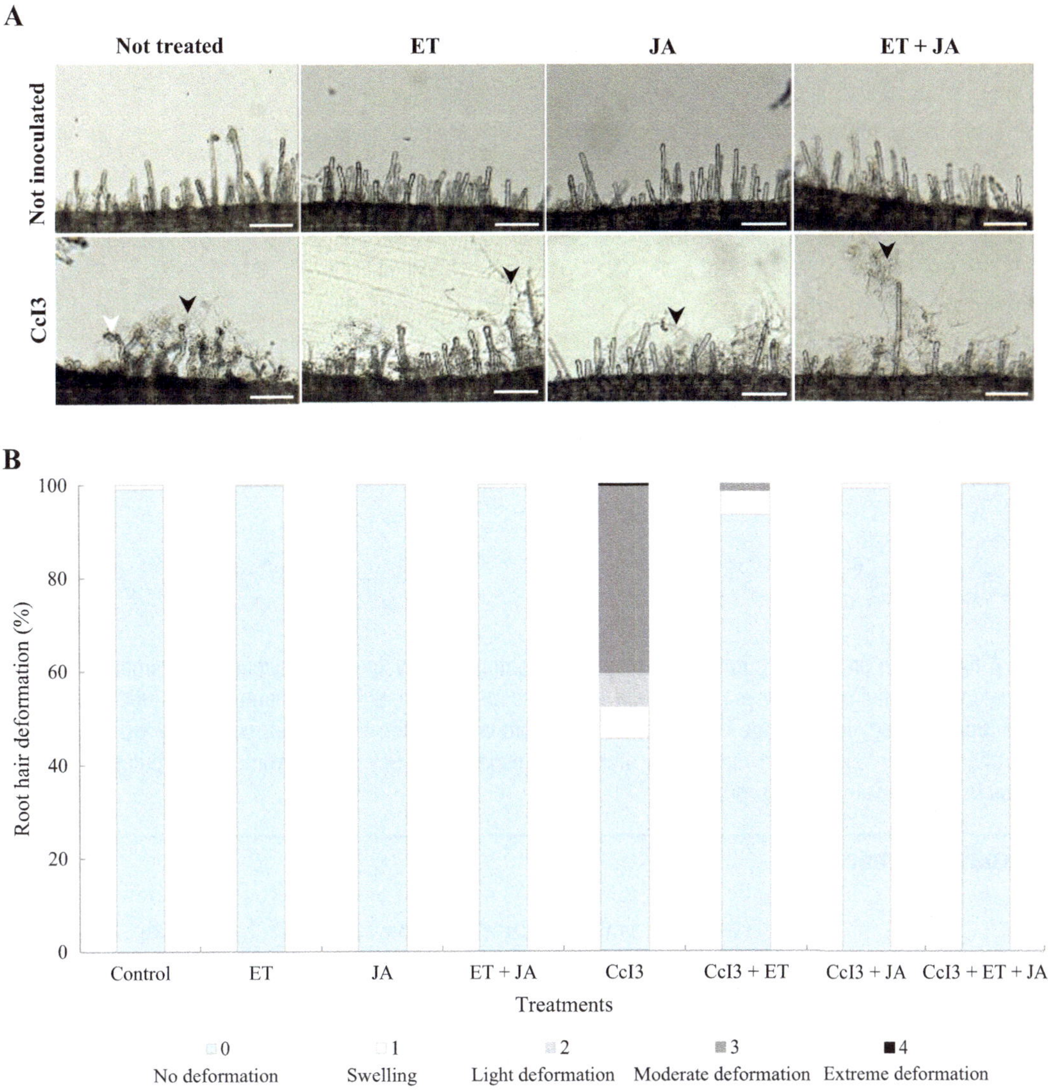

Fig. 3 Effects of ethylene and jasmonate on *C. glauca* root hair deformation. (**a**) Root hairs under control, ethylene, and/or jasmonate pretreated conditions from uninoculated and inoculated *C. glauca* plants, observed 48 h after inoculation. White arrow indicates moderate deformation and black arrows *Frankia* hyphae (Bars, 100 μm). (**b**) Quantification of root hair deformation showing the proportion of deformed root hairs in short lateral roots, 48 h after inoculation with *Frankia* strain CcI3. For each treatment, five plants were used, and three lateral roots were observed per plant

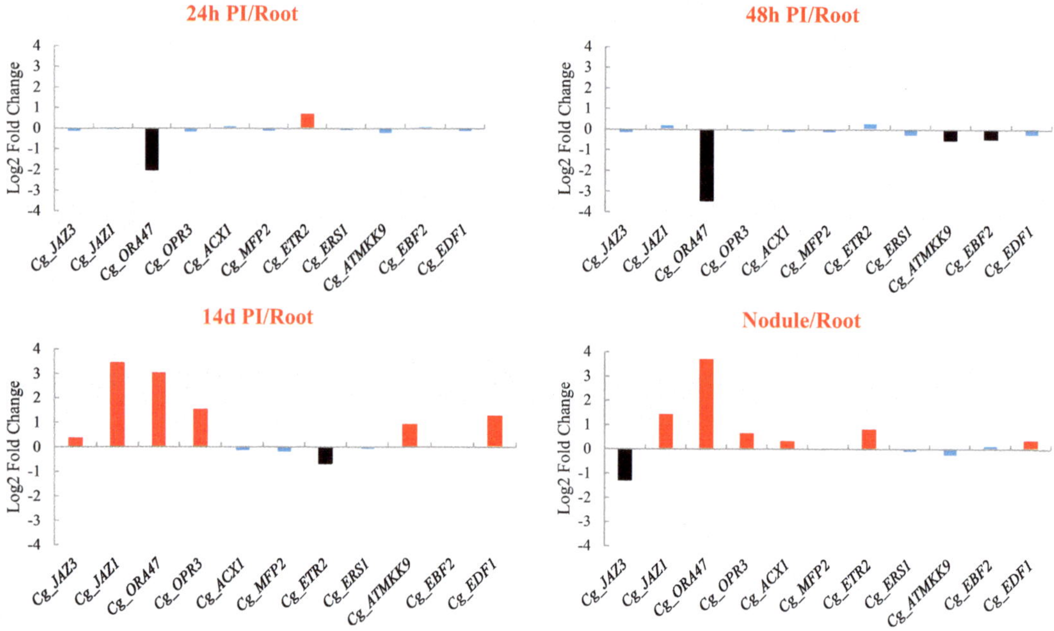

Fig. 4 Expression profiles of putative ortholog genes potentially involved in ethylene and jasmonate biosynthesis and/or signaling pathways in *C. glauca* inoculated roots and nodules compared to non-inoculated roots. Bars represent $\log_2$ fold change in inoculated roots compared to non-inoculated roots (padj ≤ 0.05 and $\log_2$ fold change + or −).Red and black bars represent, respectively, up- and downregulated putative ortholog genes in *C. glauca* inoculated roots

Acknowledgments

This work was supported, in part, by the Institut de Recherche pour le Développement (IRD) and the Agence Universitaire de la Francophonie (AUF) through the interregional doctoral college in food and plant biotechnology (CD-BIOVEGAGRO) and grants from the USDA National Institute of Food and Agriculture (Hatch 022821) and USDA AFRI (A1151 2014-03765). M. Ngom was supported by the Ministère de l'Enseignement Supérieur et de la Recherche du Sénégal, national grant (MERS), the Word Federation of Scientists, research allowance (WFS), and the ARTS, PhD grant (IRD).

References

1. Davies PJ (2010) The plant hormones: their nature, occurrence, and functions. In: Plant hormones. Springer, New York, pp 1–15
2. Bari R, Jones JD (2009) Role of plant hormones in plant defence responses. Plant Mol Biol 69:473–488
3. Peleg Z, Blumwald E (2011) Hormone balance and abiotic stress tolerance in crop plants. Curr Opin Plant Biol 14:290–295
4. Verma V, Ravindran P, Kumar PP (2016) Plant hormone-mediated regulation of stress responses. BMC Plant Biol 16:86
5. Klessig DF, Malamy J (1994) The salicylic acid signal in plants. Plant Mol Biol 26:1439–1458
6. Raskin I (1992) Role of salicylic acid in plants. Annu Rev Plant Biol 43:439–463
7. Wasternack C, Hause B (2013) Jasmonates: biosynthesis, perception, signal transduction

and action in plant stress response, growth and development. An update to the 2007 review in annals of botany. Ann Bot 111:1021–1058
8. Gamalero E, Glick BR (2012) Ethylene and abiotic stress tolerance in plants. In: Environmental adaptations and stress tolerance of plants in the era of climate change. Springer, New York, pp 395–412
9. Yan Y, Borrego E, Kolomiets MV (2013) Jasmonate biosynthesis, perception and function in plant development and stress responses. In: Lipid metabolism. IntechOpen, London
10. Gundlach H, Müller MJ, Kutchan TM, Zenk MH (1992) Jasmonic acid is a signal transducer in elicitor-induced plant cell cultures. Proc Natl Acad Sci U S A 89:2389–2393
11. Kapoor R (2008) Induced resistance in mycorrhizal tomato is correlated to concentration of jasmonic acid. Online J Biol Sci 8:49–56
12. Vijayan P, Shockey J, Lévesque CA, Cook RJ (1998) A role for jasmonate in pathogen defense of Arabidopsis. Proc Natl Acad Sci U S A 95:7209–7214
13. Feng Y-J, Wang J-W, Luo S-M (2007) Effects of exogenous jasmonic acid on concentrations of direct-defense chemicals and expression of related genes in bt (*Bacillus thuringiensis*) corn (*Zea mays*). Agric Sci China 6:1456–1462
14. Chen H-Y, Hsieh E-J, Cheng M-C, Chen C-Y, Hwang S-Y, Lin T-P (2016) ORA47 (octadecanoid-responsive AP2/ERF-domain transcription factor 47) regulates jasmonic acid and abscisic acid biosynthesis and signaling through binding to a novel cis-element. New Phytol 211:599–613
15. Pauwels L, Goossens A (2011) The JAZ proteins: a crucial interface in the jasmonate signaling cascade. Plant Cell 23:3089–3100
16. Schaller F, Biesgen C, Müssig C, Altmann T, Weiler EW (2000) 12-Oxophytodienoate reductase 3 (OPR3) is the isoenzyme involved in jasmonate biosynthesis. Planta 210:979–984
17. Cacas JL, Pré M, Pizot M, Cissoko M, Diedhiou I, Jalloul A, Doumas P, Nicole M, Champion A (2017) GhERF-IIb3 regulates the accumulation of jasmonate and leads to enhanced cotton resistance to blight disease. Molecular Plant Pathology 18:825–836
18. Wang KL-C, Li H, Ecker JR (2002) Ethylene biosynthesis and signaling networks. Plant Cell 14:S131–S151
19. Van Loon LC, Geraats BP, Linthorst HJ (2006) Ethylene as a modulator of disease resistance in plants. Trends Plant Sci 11:184–191
20. Lorenzo O, Chico JM, Sánchez-Serrano JJ, Solano R (2004) JASMONATE-INSENSITIVE1 encodes a MYC transcription factor essential to discriminate between different jasmonate-regulated defense responses in Arabidopsis. Plant Cell 16:1938–1950
21. Anderson JP, Badruzsaufari E, Schenk PM, Manners JM, Desmond OJ, Ehlert C, Maclean DJ, Ebert PR, Kazan K (2004) Antagonistic interaction between abscisic acid and jasmonate-ethylene signaling pathways modulates defense gene expression and disease resistance in Arabidopsis. Plant Cell 16:3460–3479
22. Oldroyd GE, Engstrom EM, Long SR (2001) Ethylene inhibits the nod factor signal transduction pathway of *Medicago truncatula*. Plant Cell 13:1835–1849
23. Sun J, Cardoza V, Mitchell DM, Bright L, Oldroyd G, Harris JM (2006) Crosstalk between jasmonic acid, ethylene and Nod factor signaling allows integration of diverse inputs for regulation of nodulation. Plant J 46:961–970
24. Reid D, Liu H, Kelly S, Kawaharada Y, Mun T, Andersen SU, Desbrosses G, Stougaard J (2018) Dynamics of ethylene production in response to compatible Nod factor. Plant Physiol 176:1764–1772
25. Van Spronsen PC, Tak T, Rood AM, van Brussel AA, Kijne JW, Boot KJ (2003) Salicylic acid inhibits indeterminate-type nodulation but not determinate-type nodulation. Mol Plant-Microbe Interact 16:83–91
26. Tromas A, Parizot B, Diagne N, Champion A, Hocher V, Cissoko M, Crabos A, Prodjinoto H, Lahouze B, Bogusz D et al (2012) Heart of endosymbioses: transcriptomics reveals a conserved genetic program among arbuscular mycorrhizal, actinorhizal and legume-rhizobial symbioses. PLoS One 7: e44742
27. Diédhiou I, Tromas A, Cissoko M, Gray K, Parizot B, Crabos A, Alloisio N, Fournier P, Carro L, Svistoonoff S (2014) Identification of potential transcriptional regulators of actinorhizal symbioses in Casuarina glauca and *Alnus glutinosa*. BMC Plant Biol 14:342
28. Zhong C, Mansour S, Nambiar-Veetil M, Bogusz D, Franche C (2013) *Casuarina glauca*: a model tree for basic research in actinorhizal symbiosis. J Biosci 38:815–823
29. Hocher V, Alloisio N, Auguy F, Fournier P, Doumas P, Pujic P, Gherbi H, Queiroux C, Da Silva C, Wincker P et al (2011) Transcriptomics of actinorhizal symbioses reveals homologs of the whole common symbiotic signaling cascade. Plant Physiol 156:700–711
30. Zhang Z, Lopez MF, Torrey JG (1984) A comparison of cultural characteristics and

infectivity of Frankia isolates from root nodules of Casuarina species. In: Frankia Symbioses. Springer, New York, pp 79–90

31. Murry MA, Fontaine MS, Torrey JG (1984) Growth kinetics and nitrogenase induction in Frankia sp. HFPArI 3 grown in batch culture. In: Frankia symbioses. Springer, New York, pp 61–78
32. Broughton WJ, Dilworth MJ (1971) Control of leghaemoglobin synthesis in snake beans. Biochem J 125:1075–1080
33. Hoagland DR, Arnon DI et al (1950) The water-culture method for growing plants without soil. Circ Calif Agric Exp Stn 347:32
34. Lamesch P, Berardini TZ, Li D, Swarbreck D, Wilks C, Sasidharan R, Muller R, Dreher K, Alexander DL, Garcia-Hernandez M (2011) The Arabidopsis information resource (TAIR): improved gene annotation and new tools. Nucleic Acids Res 40:D1202–D1210
35. Gonzales MD, Archuleta E, Farmer A, Gajendran K, Grant D, Shoemaker R, Beavis WD, Waugh ME (2005) The legume information system (LIS): an integrated information resource for comparative legume biology. Nucleic Acids Res 33:D660–D665
36. Clavijo F, Diedhiou I, Vaissayre V, Brottier L, Acolatse J, Moukouanga D, Crabos A, Auguy F, Franche C, Gherbi H et al (2015) The Casuarina NIN gene is transcriptionally activated throughout Frankia root infection as well as in response to bacterial diffusible signals. New Phytol 208:887–903
37. Ngom M, Gray K, Diagne N, Oshone R, Fardoux J, Gherbi H, Hocher V, SVISTOONOFF S, Laplaze L, Tisa L et al (2016) Symbiotic performance of diverse Frankia strains on salt-stressed Casuarina glauca and *Casuarina equisetifolia* plants. Front Plant Sci 7:1331
38. Ngom M, Diagne N, Laplaze L, Champion A, Sy MO (2016) Symbiotic ability of diverse Frankia strains on *Casuarina glauca* plants in hydroponic conditions. Symbiosis 70:79–86
39. Grennan AK (2006) Genevestigator. Facilitating web-based gene-expression analysis. Plant Physiol 141:1164–1166

Part III

Metabolism and Signaling

Chapter 10

Crystallization of a Complex Between MYC and Jas Motif

Feng Zhang, Sheng Yang He, and Karsten Melcher

Abstract

In the jasmonate signaling pathway, a region of 17 amino acids within the Jas motif of JAZ proteins and a conserved region within the N-terminus of MYC proteins are sufficient for JAZ–MYC interactions. Crystal structures of Jas–MYC complexes have revealed the structural basis of this important interaction. Here, we describe methods of cloning, expression, and purification of MYC N-terminal proteins and their co-crystallization with Jas motif peptides.

Key words Jasmonate, MYC transcription factor, JAZ transcription repressor, Jas motif, Crystallization

1 Introduction

The plant hormone jasmonate (JA) functions as a key plant defense signal and an important regulator of plant growth and development [1, 2]. In the stress-free stage, key mediators of jasmonate signaling include MYC transcription factors, which are restrained by a group of nuclear transcriptional repressor called JASMONATE ZIM DOMAIN (JAZ) proteins [2, 3]. JAZ proteins repress MYC target gene expression at two levels. First, they directly block the activation function of MYC transcription factors by binding and masking of their transcriptional activation domains [4]. Second, they actively repress MYC target gene expression by recruiting members of the TOPLESS (TPL) or TPL-related protein (TPR) class of corepressor. Recruitment can occur either by directly interacting with TPL through ethylene response factor–associated amphiphilic repression (EAR) motifs located at the N-terminus of JAZ proteins or by indirectly binding the EAR motif-containing adaptor protein NINJA through the ZIM domain located at the center of JAZ proteins [5–8]. TPL or TPR proteins then recruit chromatin-modifying proteins and repressive histone deacetylases to repress MYC gene expression [7, 9]. In contrast, in response to wounding, herbivorous insects and necrotrophic pathogens, as well as certain

Antony Champion and Laurent Laplaze (eds.), *Jasmonate in Plant Biology: Methods and Protocols*, Methods in Molecular Biology, vol. 2085, https://doi.org/10.1007/978-1-0716-0142-6_10, © Springer Science+Business Media, LLC, part of Springer Nature 2020

developmental or growth cues, the active form of jasmonate—jasmonoyl isoleucine (JA-Ile)—is synthesized in the plant cytosol. JA-Ile then promotes the formation of a coreceptor complex, consisting of JAZ, JA-Ile, and coronatine insensitive 1 (COI1), an F-box protein which is a subunit of an SKP–Cullin–F-box (SCF) ubiquitin ligase complex [3]. This complex formation leads to ubiquitination and proteasome-dependent degradation of transcriptional repressor JAZ proteins, which, in turn, releases MYC proteins from transcriptional repression [1, 10].

In *Arabidopsis*, 13 genes encode JAZ proteins (including JAZ1-JAZ13). They are characterized by a highly conserved central-ZIM domain and C-terminal Jas motif [11, 12]. Through this Jas motif, JAZ proteins physically interact and inhibit the MYC family proteins, MYC2, MYC3, and MYC4 [4, 13]. MYC proteins are basic–helix–loop–helix (bHLH) transcription factors, which encompass the previously defined JAZ-interacting domain (JID), and the transcription activation domain (TAD) within the N termini of all three proteins [14–16]. Crystal structures and mutational analysis have shown that JID and TAD together form a binding groove that can be competitively occupied, either by the Jas motif for transcriptional repression or upon JA-Ile-mediated JAZ proteolysis, by the mediator subunit MED25 for transcription activation. Although JID–TAD in MYC proteins and the Jas motif in JAZ proteins are sufficient for MYC–JAZ interactions, details of these interactions at the atom level required crystallization of their complex and determined the crystal structure. Here, we describe an efficient protocol for the purification of MYC proteins, using MYC3 as an example, and their co-crystallization with Jas motif peptides.

2 Materials

2.1 Genes and Vector

1. *Arabidopsis* MYC3 coding sequence (At5g46760).
2. *Arabidopsis* JAZ9 coding sequence (At1g70700).
3. SUMO coding sequence (556484, NCBI).
4. pETDuet-1 vector (Addgene).

2.2 Cloning of pETDuet-1-His6-Sumo-MYC Expression Construct

1. Phusion High-Fidelity (HF) polymerase (New England Biology, Ipswich, MA).
2. A 5 × HF buffer (New England Biology, Ipswich, MA).
3. A 10-mM deoxynucleotide (dNTP) mixture (Invitrogen, Thermo Fisher Scientific, USA).
4. Oligonucleotide primers (10 pmol/μl).

(a) *Bam*H1-MYC3-44-ATG-Forward for cloning MYC3 (44-238): ATC*GGATCC***ATG**CAGCCTCAGTTCAACGAAGATAC.

(The sequence in italics is a *Bam*H1 restriction sites. The bold sequence is the start codon. Underlined sequences are primer-binding sites).

(b) *Not*1-MYC3-238-Reverse for cloning MYC3(44-238): ATT*GCGGCCGC***TTA**GTTGAAATTAAACAAGTTGTTAA.

(The sequence in italics is a *Not*1 restriction sites. Bold sequences are stop codon. Underlined sequences are primer-binding sites).

(c) SUMO-forward for sequencing: GTTGAAATTAAACAAGTTGTTAA.

(d) T7 term-reverse for sequencing: CTAGCATAACCCCTTGGGGCCTCTAAACGGGTCTTGAGGGGTTTTTTG.

5. Sterile dH_2O.
6. Sterile RNase/DNase-free PCR tubes (Thermo Fisher Scientific, USA).
7. Sterile 1.5-ml RNase/DNase-free microcentrifuge tubes.
8. Agarose 1™ (Amersco, Solon, OH).
9. SYBR Green 1 dye (Life Technologies, Carlsbad).
10. A 50 × TAE buffer (2 M Tris, 6.05% (v/v) acetate acid, 0.5 M sodium ethylenediaminetetraacetate (EDTA)).
11. A 10 × DNA sample loading buffer (0.25% (w/v) bromophenol blue, 0.1% (w/v) SDS, 0.25% (w/v) xylene cyanol and 30% (v/v) glycerol).
12. A 1-kb DNA ladder.
13. Gel DNA Recovery and Clean-up kit (Omega Bio-tek, Norcross, GA).
14. Restriction enzymes, 10 × HF restriction buffer (New England Biology, Ipswich, MA).
15. T4 ligase and 5 × T4 ligase buffer (NEB, Ipswich, MA).
16. pETDuet-1-His6-Sumo plasmid DNA (~100 ng/ml).
17. LB broth (Caisson, Smithfield, UT).
18. Agar (EMD Millipore, Billerica, Massachusetts).
19. LB agar plate, 20 g LB broth with 1.5% (w/w) agar, fill up with dH_2O to 1 l, sterilize at 121 °C, 15 min.
20. Sterile 15-ml plastic snap-cap culture tubes.
21. Ampicillin (Amp) 50 mg/ml, filter-sterilized in dH_2O.
22. Competent *Escherichia coli* (*E. coli*) strain DH5α (Catalog#18265017, Thermo Fisher Scientific, USA).

23. Plasmids miniprep kit (Omega Bio-tek, Norcross, GA).
24. (NanoDrop Microvolume) Spectrophotometer (Thermo Fisher Scientific, USA).

2.3 MYC Expression in and Extraction from E. coli

1. Competent *E. coli* strain BL21 (DE3) (Catalog# C600003, Thermo Fisher Scientific, USA).
2. LB broth (Caisson, Smithfield, UT).
3. LB liquid: 20 g LB broth fill with dH_2O up to 1 l, sterilize at 121 °C, 15 min.
4. Agar (EMD Millipore, Billerica, Massachusetts).
5. LB agar plate: 20-g LB broth with 1.5% agar, fill with dH_2O up to 1 l, sterilize at 121 °C, 15 min.
6. Sterile 15-ml plastic snap-cap culture tubes.
7. Sterile 250-ml glass Erlenmeyer.
8. Sterile 4-l glass Erlenmeyer.
9. Ampicillin (amp) 50 mg/ml, filter-sterilized in dH_2O.
10. A 0.5-M isopropyl β-D-1-thiogalactopyranoside (IPTG; 2.97 g IPTG, fill with dH_2O up to 50 ml, filter-sterilize; Sigma-Aldrich, Burlington, MA).
11. Wash buffer (20 mM Tris, pH 8.0, 200 mM NaCl, and 10% [v/v] glycerol).
12. Extract buffer A (20 mM Tris, pH 8.0, 200 mM NaCl, 25 mM imidazole and 10% (v/v) glycerol).
13. Extract buffer B (20 mM Tris, pH 8.0, 200 mM NaCl, 500 mM imidazole, and 10% [v/v] glycerol).
14. SUMO protease (250 units) (Thermo Fisher Scientific, USA).
15. 5-ml HisTrap HP column (GE healthcare, Thermo Fisher Scientific, USA).
16. HiLoad 26/60 Superdex 200 gel filtration column (GE healthcare, Thermo Fisher Scientific, USA).
17. Gel filtration buffer (25 mM Tris, pH 8.0, 200 mM ammonium acetate, 1 mM dithiothreitol [DTT] and 1 mM EDTA).
18. Chromatography system, e.g., AKTA pure (GE Healthcare).
19. (NanoDrop Microvolume) Spectrophotometer (Thermo Fisher Scientific, USA).

2.4 SDS Polyacrylamide Gel

1. Stacking gel stock: (0.5 M Tris–HCl, pH 6.8; 10% (w/v) SDS; Acrylamide/Bis-acrylamide (30%/0.8%, w/v); 10% (w/v) ammonium persulfate (AP); and TEMED.
2. Separating gel stock: 1.5 M Tris–HCl, pH 8.8; 10% (w/v) SDS; Acrylamide/Bis-acrylamide (30%/0.8% w/v); 10% (w/v) ammonium persulfate (AP); and TEMED.
3. Protein electrophoresis equipment (Bio-rad).

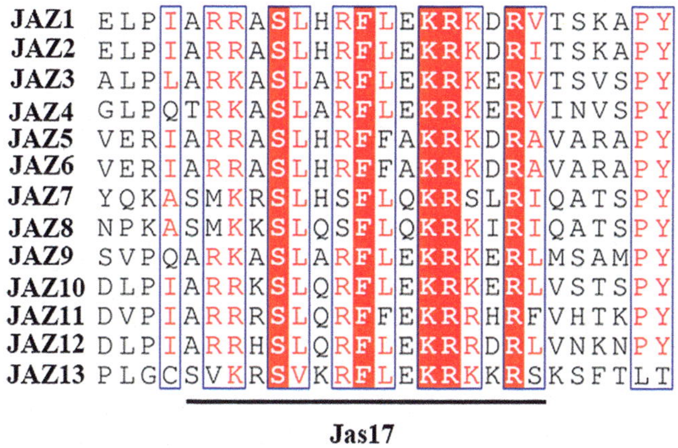

Fig. 1 Sequence alignment of the Jas motif of the 13 *Arabidopsis* JAZ proteins. Amino acids with blue frame denote conservation in all of the sequences. The underlined sequences are the region of 17 amino acids

2.5 Order Jas Motif Peptide

1. All Jas motif peptides were ordered from PEPTIDE 2.0 (Chantilly, VA).
2. NanoDrop Microvolume Spectrophotometers (Thermo Fisher Scientific, USA).
3. Have at least 5 mg peptide synthesized at a purity of $\geq$95%. Make sure all motifs are sufficient for MYC interaction first. As example, we use Jas motif peptide from JAZ9 (*see* Fig. 1).

2.6 Co-crystallization of MYC N-Terminal Protein and Jas Motif Peptide

1. Crystallization buffer (25 mM Tris, pH 8.0, 200 mM ammonium acetate, 1 mM DTT and 1 mM EDTA).
2. Crystallization kit, Index HT (Hampton Research, Aliso Viejo, CA).
3. Crystallization Screen kit, Crystal screen HT (Hampton Research, Aliso Viejo, CA).
4. Crystallization Screen kit, MembFac HT (Hampton Research, Aliso Viejo, CA).
5. Crystallization Screen kit, PEG/Ion HT (Hampton Research, Aliso Viejo, CA).
6. Crystallization Screen kit, PEGR x HT (Hampton Research, Aliso Viejo, CA).
7. Crystallization Screen kit, SaltR x HT (Hampton Research, Aliso Viejo, CA).
8. Crystallization Optimization Kit, Additive Screen (Hampton Research, Aliso Viejo, CA).
9. Crystallization Optimization Kit, Detergent Screen (Hampton Research, Aliso Viejo, CA).

10. Crystal phoenix robot (Art Robbins Instruments, Sunnyvale, CA).
11. 96-Wells sitting drop intelli-plates (Art Robbins Instruments, Sunnyvale, CA).

3 Methods

3.1 Cloning the pETDuet-1-His6-Sumo Construct

1. Design primers and introduce His6-Sumo sequence into pETDuet-1 empty vector by PCR.
2. Introduce *Bam*H1 and *Not*1 restriction sites after the His6-Sumo sequence to make pETDuet-1-His6-Sumo expression vector (*see* **Note 1**).
3. Design primers for amplification of MYC3(44-238) from MYC3 template DNA by PCR (*see* Fig. 2, **Note 2**). Primers need to contain 3′ regions matching the start and end of the MYC3 fragment and 5′ regions containing restriction sites for cloning into the multiple cloning site of pETDuet-1-His6-Sumo expression vector (here, *Bam*H1 and *Not*1) (*see* **Notes 3** and **4**).

   Examples:

   (a) *Bam*H1-MYC3-44-ATG-Forward. (The reading frame starts from codon 44 with an introduced ATG as the start codon.)

   (b) *Not*1-MYC3-238-reverse. (The reading frame stops as codon 238 with TTA as the stop codon.)
4. Set up a PCR reaction in a sterile RNase/DNase-free PCR tube according to the following recipe:

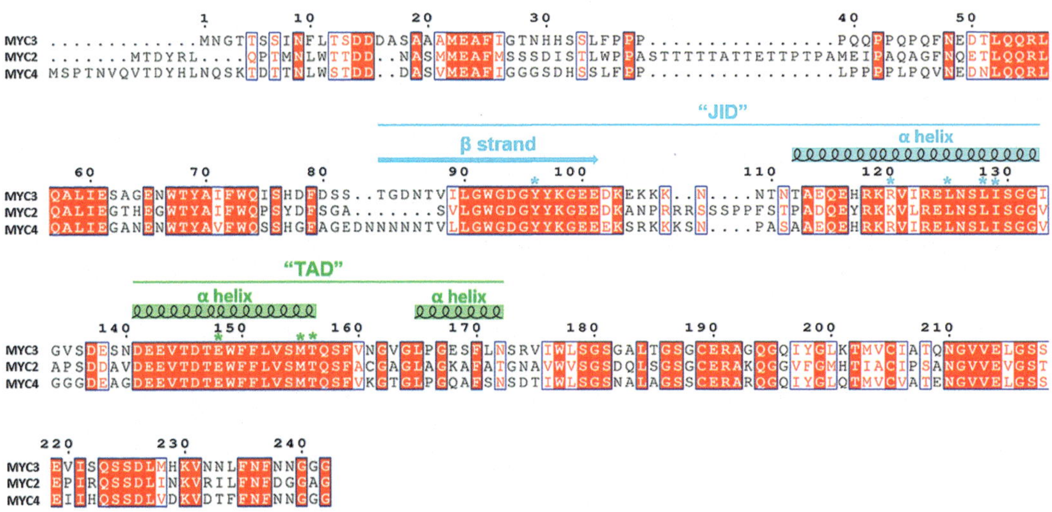

Fig. 2 Sequence alignment of Arabidopsis MYC2, MYC3, and MYC4 N termini. JID domain and TAD domain are marked with blue and green with secondary structure elements above the alignment, respectively

A 10 μl 5 × HF buffer.

A 1 μl dNTPs (10 mM).

A 0.5 μl Phusion High Fidelity polymerase.

A 1 μl MYC3 template DNA (50 ng/μl).

A 1 μl Forward primer.

A 1 μl Reverse primer.

A 35.5 μl sterile dH_2O.

5. Perform the PCR reaction in a thermocycler. The following thermocycler program is recommended:

   1 cycle: 1 min at 95 °C.

   30 cycles:

   0.5 min at 94 °C.

   0.5 min at (lowest primer T_m −3 °C).

   0.5 min/kb at 72 °C.

   (Base the extension time on the size of the PCR products).

   1 cycle: 10 min at 72 °C.

   Hold at 12 °C.

6. Separate PCR reaction by agarose gel electrophoresis and stain with SYBR Green 1 dye. The band of the amplification product should be strong and at the correct size.
7. Use the Gel DNA Recovery kit to purify the PCR product.
8. Digest 1 μl of the purified PCR product (should be about 500 ng) in a 25-μl reaction volume with the restriction enzymes that cleave within the primer sequence according to the following recipe:

   A 1 μl purified DNA.

   A 2.5 μl 10× NEB HF buffer.

   A 1 μl NEB restriction enzyme for cleavage within the forward primer.

   A 1 μl NEB restriction enzyme for cleavage within the reverse primer.

   19.5 μl dH_2O.

   Incubate the reaction for at least 3 h in 37 °C incubator.

9. Digest 500 ng of pETDuet-1-His6-Sumo expression vector in a 25-μl reaction volume with the same two restriction enzymes as used for the PCR product digest.

   A 1 μl vector plate (500 ng/μl).

   A 2.5 μl 10× NEB HF buffer.

   A 1 μl NEB internal restriction enzyme.

   A 1 μl NEB internal restriction enzyme.

A 19.5 μl dH_2O.

Incubate the reaction for at least 3 h in 37 °C incubator.

10. Purify MYC3 PCR and vector fragments with DNA Clean-up kit, elute with a 12-μl sterile dH_2O.
11. Ligate MYC3 DNA and vector with T4 DNA ligase according to the following recipe:

    A T4 DNA ligase.

    A 2 μl MYC3 DNA.

    A 0.5 μl vector.

    A 1 μl 5 × rapid ligation buffer.

    A 1 μl dH_2O.
12. Prepare LB amp plates.
13. Transform the ligation reaction into competent *E. coli* strains DH5α or Omnimax, using chemically competent heat-shock transformation protocols [17].
14. Plate one-fifth of each transformation reaction onto LB amp plates. Incubate the plates overnight at 37 °C.
15. Choose four colonies and incubate into 5-ml overnight LB amp, each. Culture overnight at 37 °C.
16. Purify the plasmid DNA from the cultures, using a Plasmid miniprep kit (e.g., from Omega).
17. Have candidate clones sequenced by commercial DNA sequencing provider, with SUMO-forward and T7term-reverse (designed in **step 4**, in Subheading 2.2) as primers to confirm correct construct and lack of PCR-induced point mutations.

3.2 MYC Expression in and Extraction from E. coli

1. Transform pETDuet-1-His6-Sumo-MYC3(44-238) plasmid into chemically competent *E. coli* strain BL21(DE3), using heat-shock transformation protocols.
2. Plate 1/5 of each transformation reaction onto LB amp plates. Incubate the plates overnight at 37 °C.
3. Pick a single colony and start 50 ml overnight LB amp culture at 37 °C.
4. Prepare 2 l LB in 4l flask (*see* **Note 3**).
5. Transfer 50 ml cultured from **step 3** into 2 l LB and culture in shaker at 30 °C to a final absorbance at 600 nm (A_{600}) of ~1.0, reduce temperature to 16 °C and incubate with 0.1 mM IPTG for 16 h (*see* **Note 5**).
6. Harvest cells in 100 ml wash buffer, then resuspend in 100 ml extract buffer A. Lyse cells by passing three times through a French press with pressure set at 1000 Pa (*see* **Notes 6** and 7).

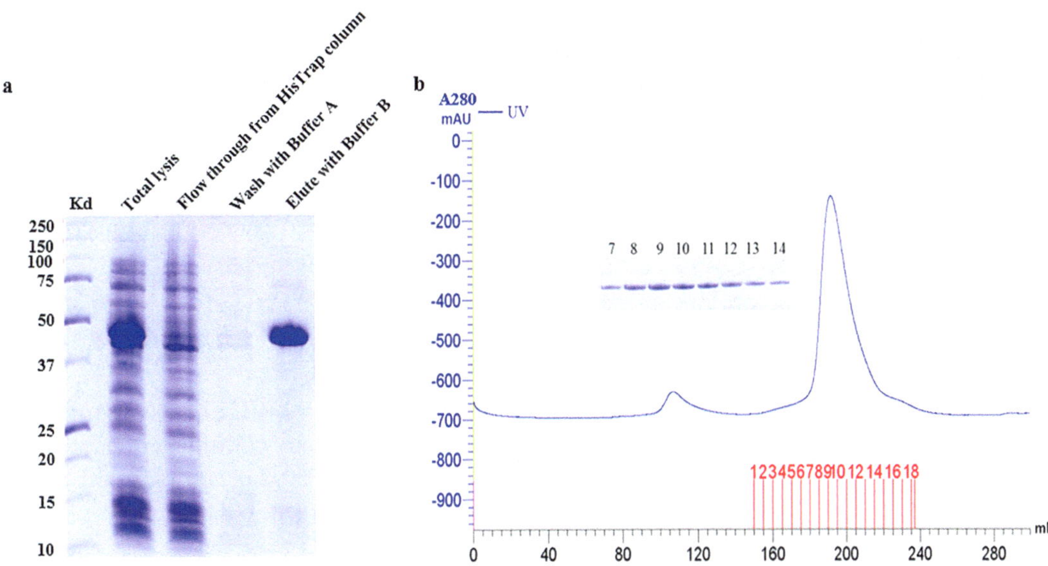

Fig. 3 Small-scale purification of His6-Sumo-MYC(44-238). (**a**) SDS-PAGE gel with 7 μl samples from different purification steps; (**b**) Size exclusion chromatogram (SEC). The red numbers indicate collected SEC fractions. Inlet: SDS-PAGE gel with 4 μl aliquots of SEC collection fractions 7–14

7. Centrifuge the lysate at 40,900 × *g* for 30 min.
8. Load the supernatant on a 5-ml HisTrap HP column.
9. Wash the column with 60 ml of extract buffer A.
10. Elute the His6-Sumo-MYC3(44-238) with 30 ml of extract buffer B.
11. Dialyze the eluted protein against extract buffer A and cleave overnight with SUMO protease at a protease:protein ratio of 1:1000 at 4 °C.
12. Remove the cleaved His6-Sumo tag by passing through a 5-ml HisTrap HP column (*see* **Note 8**).
13. Load samples from each of above steps (before induction, harvest, lysis, cleared lysate, flow through, elute, elute after cleavage, and flow-through from the second 5-ml HisTrap HP column) on SDS polyacrylamide gel using the standard protocol [18] (*see* Fig. 3a).
14. Load the flow-through from the second 5-ml HisTrap HP column on a HiLoad 26/60 Superdex 200-gel filtration column in gel filtration buffer for further purification by size exclusion chromatography.
15. Analyze fractions by SDS gel electrophoresis and concentrate purified, monomeric MYC3 protein by ultrafiltration using Amicon Ultra Centrifugation filter units (Millipore Sigma) (*see* Fig. 3b).

3.3 Co-crystallize MYC Proteins and Jas Motif Peptide

1. Calculate the Jas motif peptide concentration according to the following formula: Concentration $c = \varepsilon x A_{280}$. Calculate extinction coefficient ε by ProtParam tool (https://web.expasy.org/protparam/) and determine A_{280} (absorbance at 280 nm) using a NanoDrop Microvolume Spectrophotometer.
2. Calculate the MYC3(44-238) protein concentration by NanoDrop Microvolume Spectrophotometer and adjust the concentration to 10 mg/ml in gel filtration buffer.
3. Mix Jas motif peptide with purified MYC3(44-238) protein at a molar ratio of peptide:MYC protein of 1.5:1 and incubate on ice for 0.5 h (*see* **Note 9**).
4. Perform crystallization screening for MYC and Jas motif complex using Crystal phoenix robot and commercial crystallization screens. For each condition, mix 0.2 μl complex with 0.2 μl crystallization screens in 96-well sitting drop intelli-plate at 20 °C (*see* **Notes 10** and **11**).
5. Complex crystals of about 100 μm should appear in about 5 days (*see* **Note 12** and **13**).
6. Two crystallization optimization kits, additive screen, and detergent screen kits are recommended to further optimize the crystals obtained from above (*see* **Note 14**).
7. All crystals should be serially transferred to the well solution with 20% (v/v) ethylene glycol before flash freezing in liquid nitrogen (*see* **Note 15**).

4 Notes

1. There should be several expression vectors that are suitable for MYC N-terminal proteins. For constructs other than MYC3 (44-238), we recommend to extensively test different vectors for high yield of monodisperse MYC proteins. We have introduced the SUMO tag into the pETDuet1 expression vector to improve protein solubility, based on our previous screening for MYC N-terminal proteins.
2. Here, we used MYC3(44-238) as an example. Procedures should be very similar for other MYC proteins and other MYC constructs. We did not describe protocols for the full-length MYC proteins as we found that *E. coli* expression system is not suitable for expressing the full-length MYC proteins. The N-terminus of MYC is sufficient for JAZ interaction, and so purification and co-crystallization of MYC N-termini and Jas motif should be enough to analyze JAZ and MYC interaction in vitro.

3. There are other cloning strategies to create pETDuet-1-His6-Sumo-MYC construct. We suggest start a small-scale purification to confirm protein expression before the large-scale purification.
4. We recommend to perform a sequence alignment of MYC proteins to identify conserved N-terminal and C-terminal boundaries in MYC protein structure (*see* Fig. 2).
5. Temperature for culture of BL21 cells should be lower than 30 °C for maximal yield and solubility. We recommend expressing MYC proteins at 16 °C overnight.
6. Al buffers should be cooled to 4 °C, and all procedures during purifications should be done at 4 °C.
7. Lysis by French press leads to sample heating. Make sure the sample temperature remains below room temperature to avoid protein denaturation and degradation.
8. If the cleaved His6-Sumo tag is incompletely removed by a HisTrap column, reapply flow through one or two more times to the 5-ml HisTrap HP column. The molecular weights of His6-Sumo tag and MYC3(44-238) are very close, so that they cannot be efficiently separated by the subsequent HiLoad 26/60 Superdex 200-gel exclusion chromatography (*see* Fig. 3b).
9. The Jas motif:MYC3 ratio for co-crystallization can be increased up to 2:1, followed by incubation on ice for at least 0.5 h.
10. The setting up of crystal screen for MYC–Jas complex can be done by hand manually, though this would require more proteins, peptides, and screen buffers due to larger pipetting volumes.
11. Different crystal screen kits from other companies are also recommended to screen complex crystals.
12. The complex crystals should first appear in 2 days. However, more days are required for crystals to grow bigger. Usually, different proteins will require different time periods for crystals to appear.
13. An alternative way to co-crystallize a protein complex is to generate and crystallize a covalent fusion of the binding partners to overcome dissociation of weakly associated (low-affinity) proteins.
14. In case a specific MYC–JAZ complex has not been crystallized previously, crystals should first be optimized. For original MYC3(44-238)–Jas complex, crystals were optimized by additive and detergent screen, kits as well as by 2D grid screen

optimizing precipitant, salt, and pH value, using standard protocols [19].

15. Crystals with 20% (v/v) ethylene glycol can be stored in liquid nitrogen for a long time.

Acknowledgments

This work was supported by The Fundamental Research Funds for the Central Universities KYZ201817, by Henry Fok Young Teachers fund 160022, by Asia-hub grand Asia Hub on WEF and Agriculture grand 2017-AH-03, and by National Natural Science Foundation of China grand 31871996.

References

1. Sheard LB et al (2010) Jasmonate perception by inositol-phosphate-potentiated COI1-JAZ co-receptor. Nature 468(7322):400–405
2. Chini A et al (2007) The JAZ family of repressors is the missing link in jasmonate signalling. Nature 448(7154):666–671
3. Thines B et al (2007) JAZ repressor proteins are targets of the SCF(COI1) complex during jasmonate signalling. Nature 448 (7154):661–665
4. Zhang F et al (2015) Structural basis of JAZ-mediated repression of MYC transcription factors in jasmonate signaling. Nature 525:269–273
5. Acosta IF et al (2013) Role of NINJA in root jasmonate signaling. Proc Natl Acad Sci U S A 110(38):15473–15478
6. Pauwels L et al (2010) NINJA connects the co-repressor TOPLESS to jasmonate signalling. Nature 464(7289):788–791
7. Kazan K (2006) Negative regulation of defence and stress genes by EAR-motif-containing repressors. Trends Plant Sci 11(3):109–112
8. Ke J et al (2015) Structural basis for recognition of diverse transcriptional repressors by the TOPLESS family of corepressors. Sci Adv 1: e1500107
9. Long JA, Ohno C, Smith ZR, Meyerowitz EM (2006) TOPLESS regulates apical embryonic fate in Arabidopsis. Science 312 (5779):1520–1523
10. Yan J et al (2009) The Arabidopsis CORONATINE INSENSITIVE1 protein is a jasmonate receptor. Plant Cell 21(8):2220–2236
11. Melotto M et al (2008) A critical role of two positively charged amino acids in the Jas motif of Arabidopsis JAZ proteins in mediating coronatine- and jasmonoyl isoleucine-dependent interactions with the COI1 F-box protein. Plant J 55(6):979–988
12. Thireault C et al (2015) Repression of jasmonate signaling by a non-TIFY JAZ protein in Arabidopsis. Plant J 82(4):669–679
13. Kazan K, Manners JM (2013) MYC2: the master in action. Mol Plant 6(3):686–703
14. Chen R et al (2012) The Arabidopsis mediator subunit MED25 differentially regulates jasmonate and abscisic acid signaling through interacting with the MYC2 and ABI5 transcription factors. Plant Cell 24(7):2898–2916
15. Fernández-Calvo P et al (2011) The Arabidopsis bHLH transcription factors MYC3 and MYC4 are targets of JAZ repressors and act additively with MYC2 in the activation of jasmonate responses. Plant Cell 23(2):701–715
16. Cevik V et al (2012) MEDIATOR25 acts as an integrative hub for the regulation of jasmonate-responsive gene expression in Arabidopsis. Plant Physiol 160(1):541–555
17. Sambrook J, Fritsch EF, Maniatis T (1989) Molecular cloning: a laboratory manual, 2nd edn. Cold Spring Harbor Laboratory Press, New York
18. Kurien BT, Scofield RH (2015) Multiple Immunoblots by passive diffusion of proteins from a single SDS-PAGE gel. Methods Mol Biol 1312:77–86
19. Cudney R, Patel S, Weisgraber K, Newhouse Y, McPherson A (1994) Screening and optimization strategies for macromolecular crystal growth. Acta Crystallogr D Biol Crystallogr 50:414–423

Chapter 11

A Fluorescence Anisotropy-Based Comprehensive Method for the In Vitro Screening of COI1–JAZs Agonists and Antagonists

Yousuke Takaoka, Kengo Hayashi, Kaho Suzuki, Ika Nurul Azizah, and Minoru Ueda

Abstract

The phytohormone (+)-7-iso-jasmonoyl-L-isoleucine (JA-Ile) causes protein–protein interactions (PPI) between F-box Protein CORONATINE INSENSITIVE 1 (COI1) and JASMONATE ZIM DOMAIN (JAZ) transcriptional repressor. A total of 13 *JAZ* subtypes are encoded in the genome of *Arabidopsis thaliana*; however, their genetic redundancy obfuscates the individual function of each *JAZ*. One approach to decipher this redundant signaling network is chemical genetics, using small molecules specific to individual JAZ subtype, for which a reliable and high-throughput screening system of the ligands for all combinations of COI1–JAZs would be indispensable. In this chapter, we describe a fluorescence anisotropy-based quantitative screening system for the ligands of COI1–JAZ co-receptors. Our method is applicable to agonists and antagonists of the COI1–JAZs.

Key words COI1-JAZ co-receptor, Ligand screening, Affinity, Fluorescence anisotropy, JAZ degron peptide, Protein–protein interaction

1 Introduction

Recent advances in plant hormone biology have resulted in the identification of almost all plant hormone receptors, including TIR1-AUX/IAA for auxin, GID1-DELLA for gibberellin, and COI1-JAZ for jasmonoyl L-isoleucine (JA-Ile), and so on [1]. JA-Ile interacts with COI1–JAZ co-receptor, causing a protein–protein interaction (PPI) between F-box Protein CORONATINE INSENSITIVE 1 (COI1) and JASMONATE ZIM DOMAIN (JAZ) transcriptional repressor protein to induce myriad of JA responses, such as defense responses against the attacks of herbivores and necrotrophic pathogens; growth inhibition; and secondary metabolite production [2–4].

Antony Champion and Laurent Laplaze (eds.), *Jasmonate in Plant Biology: Methods and Protocols*, Methods in Molecular Biology, vol. 2085, https://doi.org/10.1007/978-1-0716-0142-6_11,

JAZ subtypes always function redundantly and compensate each other. In *Arabidopsis thaliana*, JAZ repressor has 13 subtypes, whereas only one COI1 is encoded in the genome. Surprisingly, no remarkable phenotype was found in *jazQ* mutant, in which five *JAZ* genes (JAZ*1/3/4/9/10)* are impaired [5]. Functional redundancy among the members of such a multigene family will often hinder the genetic analysis on the contribution of individual *JAZ* genes in jasmonate responses. However, chemical biology offers a promising solution to these difficulties [6, 7]. For example, a subtype-specific small molecular agonist–antagonist of COI1–JAZ is expected to be a powerful chemical tool with which to study the complex signaling network of jasmonate. However, to date, only two rationally designed agonists/antagonists have been reported on the COI1–JAZ system, and reliable and simple methods for their screening are needed if such a development is to progress [8, 9].

Five different methods have been reported for evaluating the affinity between plant hormone receptors and their ligands: yeast-two-hybrid system (Y2H) [10], pull-down experiments (PD) [2], Surface Plasmon Resonance (SPR) [11], isothermal titration calorimetry (ITC) [11], and differential scanning fluorimetry (DSF) [12]. In principle, all of these methods can be used for the agonist/antagonist screening of plant hormone receptor, including COI1–JAZ. However, in practice, none of them are of practical use for screening: Y2H is unreliable; PD/ITC/DSF requires high concentration of the protein samples, and SPR is expensive [13].

Here, we introduce fluorescence anisotropy (FA) for the evaluation of the affinity between COI1–JAZ and small molecule ligands [14]. FA has been widely used in vitro to evaluate interaction between molecules, such as ligand–receptor and protein–protein interactions (PPI); and use of either FA or the fluorescent polarization assay is reported in the top 51% of all cases in the PPI inhibitor database [13, 15]. The degree of FA exhibited by a molecule depends on how quickly it rotates, which, in turn, is perturbed by binding between the molecule of interest and a larger protein. FA measurements enable the quantitative evaluation of the affinity between small molecules and ligands, from which a K_d value can be inferred. FA measurements can be made quickly, without the need for specialized equipment, and, therefore, FA is well suited to be used in high-throughput screening. The preparation of the Fl-conjugated JAZ short peptide and GST-tag-fused COI1, the FA titration experiments of the agonists/antagonists, and the plate reader–based assay protocol (Fig. 1) are all described.

Fluorescein-conjugated JAZ short peptides were designed with reference to the previously reported crystal structure of COI1–JA–Ile–JAZ1 complex [8, 16]. Each JAZ short peptide is composed of 27 amino acids, including Jas motif, and it has sufficient affinity to COI1 to form a stable complex with it, in the presence of the agonist (**1** or coronatine [COR, **2**] as a structural mimic of **1**).

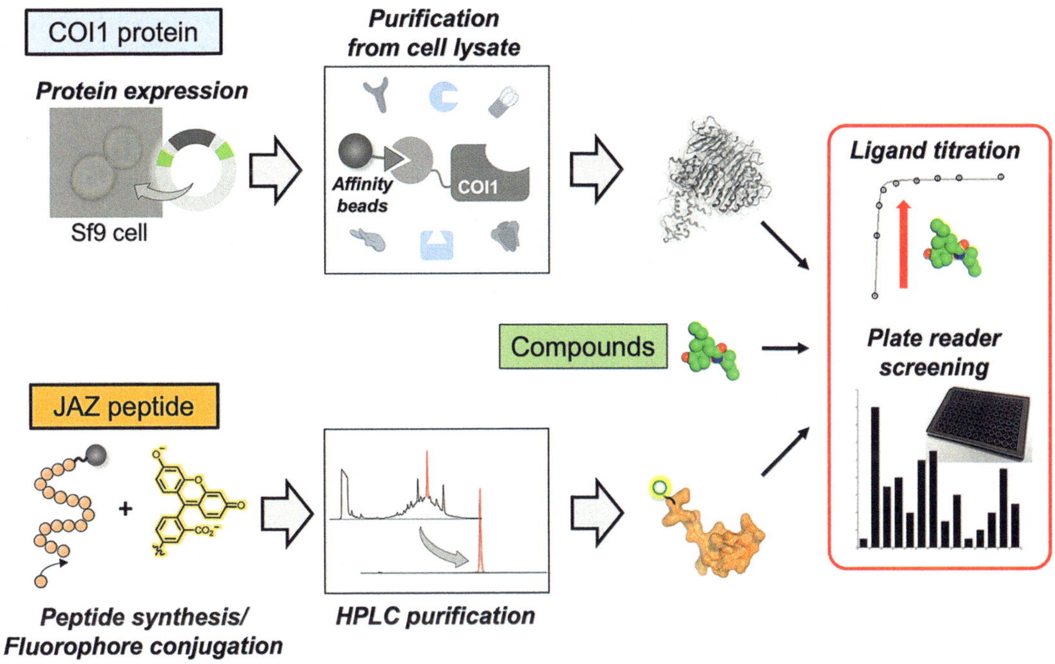

Fig. 1 Schematic of the work-flow for the FA-based binding assay system. GST-tagged COI1 protein was expressed with insect cells (Sf9)-baculovirus system, and purified from cell lysate with affinity purification (*see* Subheadings 3.1 and 3.2). Fluorescein-conjugated JAZ peptide was synthesized with Fmoc-based solid phase synthesis, and purified with HPLC (*see* Subheading 3.3). The FA-based binding assay was performed using prepared COI1 and JAZ with test compounds (*see* Subheadings 3.4–3.6)

Fluorescein (Fl), a commercially available green fluorescent molecule, was conjugated to each JAZ peptide in the N terminus of the peptide in a condensation reaction. The molecular weight of the Fl-conjugated JAZ peptide (about 3.5 kDa) was significantly less than the corresponding GST-COI1–JA–Ile–Fl–JAZ co-receptor complex (about 100 kDa), and therefore a remarkable increase in FA was expected upon ligand-mediated formation of the ternary complex, in a dose-dependent manner. This design can be applied to almost all JAZ subtype, including JAZ1–4, 5/6 (these two subtypes have identical amino acid sequence in the Jas motif), 9–12, and JAZ 13 as a negative control due to lack of the typical Jas motif [17, 18]. In contrast, jasmonic acid (JA) or *ent*COR cannot induce co-receptor complex formation and can therefore be used as a negative control. Thus, the FA-based method can specifically detect the agonist-induced PPI between COI1 and JAZ. From the saturation curve of the FA change ($r/r_0 - 1$) ratio, the apparent K_d value of the ligands can be calculated, and therefore the first quantitative evaluation of the affinities between endogenous JA derivatives and almost all the combinations of COI1–JAZs has been successfully accomplished. Our FA method can be expanded to evaluate antagonists for COI1–JAZ co-receptor by monitoring the disassembly of the GST-COI1–JA-Ile–Fl-JAZ complex. In addition, due to the high signal-to-noise ratio, our FA-based method can be performed on a microplate reader. This is a major advantage of our method for practical use. Our in vitro method has another major advantage: it can be applied for the COI1–JAZ co-receptor system of various plant species, not only of *A. thaliana* but also of commercially valuable crops.

2 Materials

2.1 Preparation of GST-COI1 and the Determination of the Effective Concentration (See Subheadings 3.1 and 3.2 and Notes 1–7)

1. Plasmids: Plasmids of GST-fused COI1 (pFB-GTE-COI1) and ASK1 (pFB-HTB-ASK1) were obtained from Addgene (https://www.addgene.org/).
2. Insect cells: Insect cells (Sf9) were cultured in ESF921 (Expression Systems LLC, Davis, CA), containing 2% vol FBS, 0.5% vol penicillin-streptomycin (PS) and 0.5 μg/mL amphotericin B at 27 °C, 125 rpm using BR-180LF (TAITEC Co. Ltd., Tokyo, Japan).
3. Lysis buffer: 20 mM Tris–HCl (pH 8.0), 200 mM NaCl, 10% vol glycerol, 10 mM 2-mercaptoethanol, and 10% vol cOmplete™ EDTA-Free. The solution was prepared according to the manufacturer's instructions.
4. Elution buffer: 50 mM Tris–HCl (pH 7.8), 100 mM NaCl, 10% vol glycerol, 0.1% vol Tween20, and 10 mM reduced glutathione.
5. Sample buffer for SDS-PAGE: 0.125 M Tris–HCl (pH 6.8), 4% w/v SDS, 20% w/v Glycerol, 0.002% w/v bromophenol blue, and 10% vol 2-mercaptoethanol (FUJIFILM Wako Pure Chemical Industries Ltd., Japan).
6. COI1 buffer for the pull-down and fluorescence anisotropy assays: 50 mM Tris–HCl (pH 7.8), 100 mM NaCl, 10% vol glycerol, and 0.1% vol Tween20, 20 mM 2-mercaptoethanol.
7. Affinity column: Glutathione Sepharose 4B (GE Healthcare UK, Ltd., UK.).
8. Bradford assay: Protein Assay Coomassie Brilliant Blue (CBB) Solution (nakalai tesque, Inc., Japan) can be used for this assay according to the manufacture information. BSA solution was used as standard protein (0.2–1.0 μg/μL, Thermo Fisher Scientific, Ltd., US).
9. Western blotting analyses: PVDF membrane, anti-GST antibody-HRP conjugate and ECL Prime western blotting detection kit (GE Healthcare, Ltd., UK).

2.2 Synthesis of Fl-Conjugated JAZ Peptide

1. Amino acids, chemicals, and solvents: 9-fluorenylmethyloxycarbonyl (Fmoc)-protected amino acid, *N*,*N*′-dicyclohexylcarbodiimide (DCC), *N*,*N*,*N*′,*N*′-tetramethyl-*O*-(1*H*-benzotriazol-1-yl)uronium hexafluorophosphate (HBTU), 1-hydroxybenzotriazole monohydrate ($HOBt{\cdot}H_2O$), *N*,*N*-diisopropylethylamine (DIPEA), pyridine, piperidine, *N*,*N*-dimethyl-4-aminopyridine (DMAP), *N*,*N*-dimethylformamide (DMF), dichloromethane (DCM), trifluoroacetic acid (TFA), and Et_2O were used without any further purification.
2. Solid-phase beads: NovaSyn® TGA resin (90 μm, 0.20–0.30 mmol/g).
3. Peptide synthesis: Initiator$^+$ Alstra (Biotage Ltd., North Carolina, US).
4. HPLC purification: PU-4180 plus with UV-4075 and MD-4010 detectors (JASCO, Tokyo, Japan).
5. Determination of the concentration of Fl-conjugated peptides: UV-2600 spectrophotometer (Shimadzu, Kyoto, Japan).

2.3 FA Measurement Using Cuvette

1. Measurement buffer: COI1 buffer (Subheading 2.1, **step 6**) containing 100 nM D-*myo*-inositol 1,2,4,5,6-pentakis-phosphate (IP5: this product was purchased from Sigma-Aldrich (USA). The sodium salt of IP5 is also available in Santa Cruz (USA) or Cayman Chemical (USA).
2. Spectrofluorometer: FP-8500 (JASCO, Tokyo, Japan).
3. Quartz cell (3 × 3 mm, 50 μL).
4. Ligand stock solution: Each ligand was dissolved in dimethyl sulfoxide (DMSO) or aqueous 20% EtOH solution.

2.4 FA Measurement Using Plate Reader

1. Measurement buffer: COI1 buffer (Subheading 2.1, **step 6**) containing 50 nM IP5.
2. Plate reader: SpectraMax-M5 (Molecular devices, LLC, U.S.).
3. 96-well plate: 96-well black clear-bottomed plate (100 μL/1 well, CORNING).

3 Methods

3.1 Preparation of GST-COI1 (Fig. 2)

1. Transform pFB-GTE-COI1 and pFB-HTB-ASK1 plasmids to MAX Efficiency DH10Bac Competent Cells (Thermo Fisher Scientific Ltd., US). Prepare the P0 virus according to the manufacturer's information (Thermo Fisher Scientific Ltd., https://tools.thermofisher.com/content/sfs/manuals/bevtest.pdf).

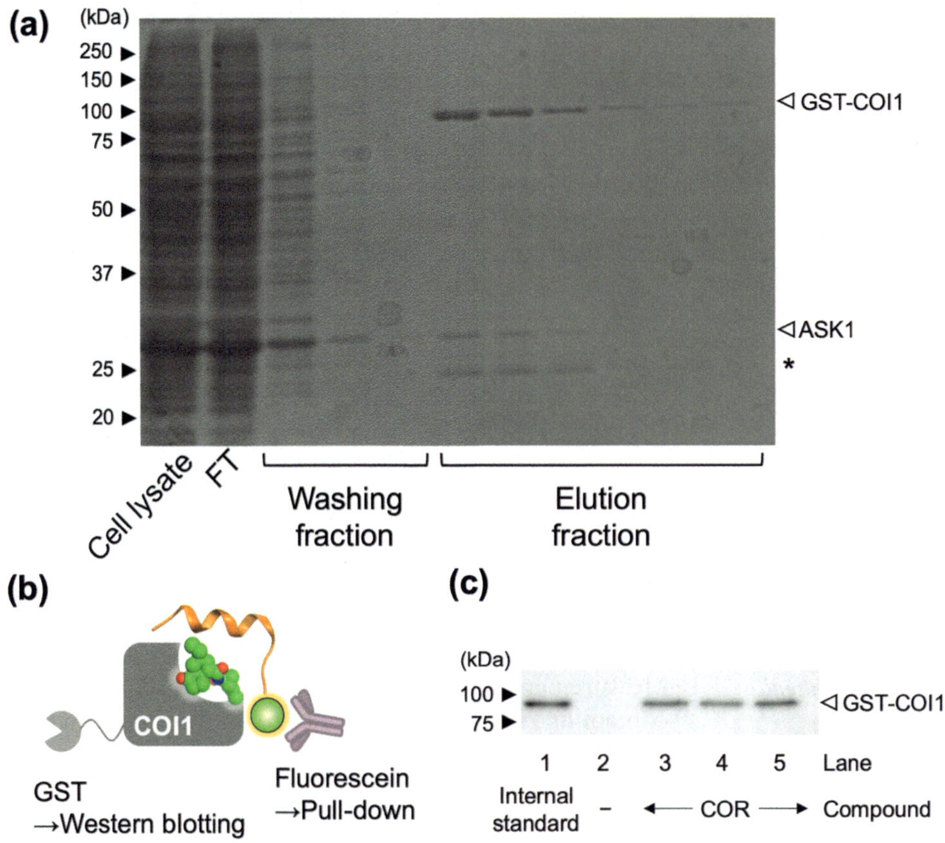

Fig. 2 Preparation of GST-COI1. (**a**) CBB stained gel subjected with each fraction. GST-COI1 and co-expressed ASK1 are shown as a triangle (asterisk indicate unknown contaminant protein). (**b**) Schematic image of the pull-down assay with GST-COI1 and fluorescein-conjugated JAZ. Fluorescein was used as epitope tag for pull-down with anti-Fluorescein antibody, and GST was used for detection tag with anti-GST antibody. (**c**) Representative results the pull-down assay for determination of the effective concentration of GST-COI1

2. Add Sf9 cells (1.0×10^6 cells/mL) in 50 mL of ESF921 containing 2% vol FBS and 0.5% vol PS to P0 virus (500 μL) and culture for 4 days at 27 °C, 125 rpm.
3. Centrifuge the cell suspension ($3200 \times g$, 4 °C, 5 min) and collect the supernatant as P1 virus (*see* **Notes 2**).
4. Measure the titer of the prepared P1 virus using the baculoQUANT™ All-In-One virus Extraction & Titration Kit (Oxford Expression Technologies Ltd., Oxford, UK).
5. Add Sf9 cells (1.0×10^6 cells/mL) in 450 mL ESF921 containing 2% vol FBS and 0.5% vol PS to P1 virus (adjust multiplicity of infection [MOI] to 0.1) and culture for 4 days at 27 °C, 125 rpm.

6. After culturing, centrifuge the suspension of the cells (3200 × *g*, 4 °C, 5 min) and collect the supernatant as P2 virus (*see* **Notes 2**).
7. Titer the prepared P2 virus, using the baculoQUANT™ All-In-One virus Extraction & Titration Kit (Oxford Expression Technologies Ltd., Oxford, UK).
8. Co-transfect Sf9 cells (1.0 × 10^6 cells/mL) in 900 mL ESF921, containing 2% vol FBS, 0.5% vol PS, and 0.5 μg/mL amphotericin B with P2 virus of COI1 and ASK1 (MOI was adjusted to 5.0) and culture for 72 h at 27 °C, 125 rpm.
9. Collect and centrifuge (3000 × *g*, 4 °C, 20 min) the cultured cells. Remove and resuspend the supernatant with 20 mL D-PBS (−) (the following protocols should be performed on ice).
10. Centrifuge the suspension of transfected cells (3000 × *g*, 4 °C, 10 min). Then, remove the supernatant (*see* **Notes 3**).
11. Suspend the pellet in 5.5 mL lysis buffer and homogenize by ultrasonication (10 s for three times with at least 30-s intervals).
12. Centrifuge the homogenate (15,000 × *g*, 4 °C, 30 min) and collect the supernatant (shown as the "cell lysate" lane in Fig. 2a).
13. Collect the soluble protein by ultracentrifugation (150,000 × *g*, 4 °C, 1.5 h).
14. Add the soluble protein to equilibrated Glutathione Sepharose 4B in lysis buffer slurry (GE Healthcare) and incubate for 1.5 h with a rotator at 4 °C (the volume of sepharose resin is about 1.0 mL).
15. Centrifuge this mixture (1000 × *g*, 4 °C, 1 min) and collect the supernatant as it flows through (shown as the "FT" lane in Fig. 2a).
16. Resuspend the sepharose resin in lysis buffer (5 mL; approximately equal to five column volumes) of lysis buffer, and remove the washing fraction after centrifugation (1000 × *g*, 4 °C, 1 min). Repeat this washing step three times.
17. Resuspend the washed sepharose resin in elution buffer (0.5 mL; approximately equal to 0.5 column volumes) and incubate for 20 min with rotator at 4 °C.
18. Centrifuge the mixture from above (1000 × *g*, 4 °C, 1 min), then collect the supernatant into a microtube to which proteins will not bind (e.g., PROKEEP from Watson Co. Ltd., Tokyo, Japan) as elution fraction.

19. Resuspend the sepharose resin in elution buffer (0.5 mL; approximately equal to 0.5 column volumes). Centrifuge (1000 × *g*, 4 °C, 1 min) and collect the supernatant as the second elution fraction.
20. Repeat this elution procedure four to six times (*see* **Notes 4**).

3.2 Determination of the Effective Concentration of GST-COI1 in Each Elution Fraction

1. Determination of Weight/Vol Concentration.

   Each fraction was analyzed by Bradford assay by using BSA as a standard (0.2–1.0 μg/μL).
2. Determination of the purity of GST-COI1 (Fig. 2a).

   Each elution fraction (15 μL) is mixed with 15 μL of 2× sample buffer for SDS-PAGE, and the solution is heated for 10 min at 60 °C. These samples are subjected to SDS-PAGE (Mini-Protean Tetra vertical electrophoresis cell, Bio-Rad Laboratories, Inc.) with a molecular weight marker. Gels are stained with CBB, and the purity of each fraction is calculated from band intensity of all the detectable protein bands using ImageJ 1.45S software (http://imagej.net/Welcome).
3. Determination of the effective concentration of GST-COI1 (Fig. 2b, c; **Notes 5–7**). The effective concentration of GST-COI1 is estimated by pull-down assay with COR and OregonGreen®-conjugated JAZ1 peptide [8]. Each fraction of GST-COI1 (5 nM), OregonGreen®-conjugated JAZ1 peptide (10 nM), inositol-1,2,4,5,6-pentakisphosphate (100 nM), and COR (100 nM) are mixed in COI1 buffer (350 μL), and the solution is incubated for 1 h on ice. Rabbit anti-fluorescein antibody (0.2 μL, Abcam, ab19491) is added and the mixture rotated overnight at 4 °C.
4. After rotation, combine the samples with Protein A Mag Sepharose Xtra (15 μL in 50% incubation buffer slurry, GE Healthcare, 28944006). After a 3-h rotation at 4 °C, wash the samples three times with 350 μL of COI1 buffer. Suspend the washed beads in 35 μL of sample buffer for SDS-PAGE containing 100 mM dithiothreitol and heat for 10 min at 60 °C. Subject the eluted samples to SDS-PAGE with COI1 internal standard sample (5 nM) and analyze by western blotting. Detect the bound COI1–GST using anti-GST HRP conjugate 5000-fold diluted with skimmed milk. Estimate the effective concentration from the pull-down rate of GST-COI1 calculated from internal standard sample (authentic GST-COI1). Calculate the pull-down rate for each fraction (*see* **Notes 5–7**).

3.3 Synthesis of Fl-Conjugated JAZ Peptide (Fig. 3)

1. Synthesize each JAZ peptide by microwave-assisted Fmoc-based solid phase synthesis with NovaSyn® TGA resin by using Initiator$^+$ Alstra. NovaSyn® Swell the TGA resin with

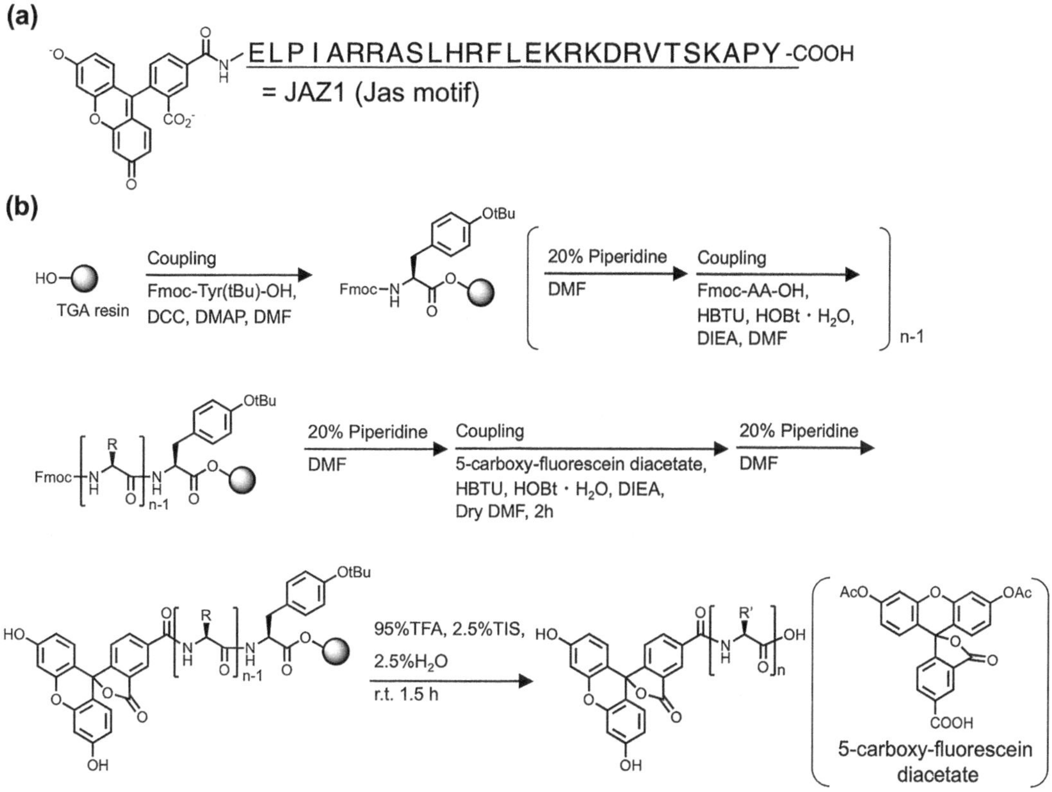

Fig. 3 Preparation of fluorescein-conjugated JAZ peptide (Fl-JAZ peptide). (**a**) Chemical structure and amino acid sequence of Fl-JAZ1 peptide. (**b**) Synthetic scheme of Fl-JAZ peptide by Fmoc-based solid phase synthesis and fluorescein conjugation. R and R′ represent protected and deprotected side chains of each amino acid

DMF for 20 min at 70 °C. Dissolve the first FMOC-protected amino acid (FMOC-Tyr(tBu)-OH for all JAZ peptides (5 eq) and DCC (5 eq) in the minimum volume of DMF (about 1 or 2 mL) and stir for 15 min at 0 °C. Add this mixture to the beads. Add DMAP (0.5 eq) and shake for 1 h. After introduction of the first amino acid, cap any unreacted resin-bond hydroxyl groups by shaking the beads in 0.25 M benzoic anhydride in 20% pyridine/DMF for 1 h. Deprotect the FMOC group by treating the beads with 20% piperidine in DMF (shake for 3 min, remove the solvent, and then shake with more 20% piperidine in DMF for 10 min). Then, treat the beads with 0.5 M FMOC-protected amino acid (3 eq) in the presence of 0.5 M HBTU (2.94 eq), 0.5 M HOBt (3 eq) and 0.5 M DIEA (3 eq) in DMF. After the reaction, remove the solvent and wash the resin three times with DMF and then three times with DCM. Repeat these processes to produce the desired peptide.

2. Mix the solid-phase JAZ peptide with 5-carboxy-fluorescein diacetate (3 eq), HBTU (5 eq), HOBt·H_2O (5 eq), and DIEA (5 eq) in dry DMF for 2 h.
3. After the reaction has completed, remove the solvent, and then wash the resin three times with DMF and three times with DCM.
4. Deprotect the peptide and elute it from the resin by stirring the beads with TFA solution containing 2.5% H_2O and 2.5% triisopropylsilane (TIS) for 1.5 h (*see* **Note 8**).
5. Precipitate the eluted peptide with cold Et_2O (about 40 mL) and collect by centrifugation (3500 × *g*, 4 °C, 5 min).
6. Obtained residue is washed with cold Et_2O twice.
7. Dry the precipitated crude peptide in vacuo, then dissolve in MeCN (0.05% TFA): H_2O (0.05% TFA) = 20:80.
8. Purify the crude peptide by RP-HPLC, using a Develosil ODS-HG-5 column (Φ4.6 × 250 mm, Nomura Chemical, Nagoya, Japan) eluting with a linear gradient (MeCN [0.05% TFA]:H_2O [0.05% TFA] = 20:80 [5 min] to 50:50 [35 min]) to afford solution of purified Fl-JAZ peptide, which should then be lyophilized.
9. After lyophilization, the purified Fl-JAZ peptide is dissolved in sterilized water to prepare the stock solution (*see* **Notes 9–11**).
10. Dilute the stock solution with aqueous 0.1 N NaOH solution for 100–500 dilution and calculate its concentration using the UV-vis absorbance at 494 nm and a molar extinction coefficient of 75,000 M^{-1} cm^{-1} [19].

3.4 Fluorescence Anisotropy Titration Experiments and Calculation of K_d Values of the Agonists of COI1–JAZ Co-Receptors (Fig. 4)

1. Measure the fluorescence anisotropy on a spectrofluorometer (Subheading 2.3, **step 2**) at room temperature with a polarized filter for both excitation and emission fluorescence signals.
2. Calculate fluorescence anisotropy values (r) using the following equation:

$$r = (I_{VV} - G \times I_{VH})/(I_{VV} + 2G \times I_{VH})$$

I_{VV} (or I_{VH}) is the fluorescence intensity observed through vertical polarizer on the excitation and vertical (or horizontal) polarization on the detected emission. G is a grating factor ($G = I_{HV}/I_{HH}$) used as a correction for the instrumental differences in detecting emitted compounds.

3. Measure the background fluorescence using 50 μL of measurement buffer in the cuvette. Background fluorescence is defined as $\lambda_{ex}/\lambda_{em}$ = 485 nm/516.5 nm.
4. Measure the correlation factor and initial fluorescence anisotropy of the solution, containing COI1–GST (100 nM) and

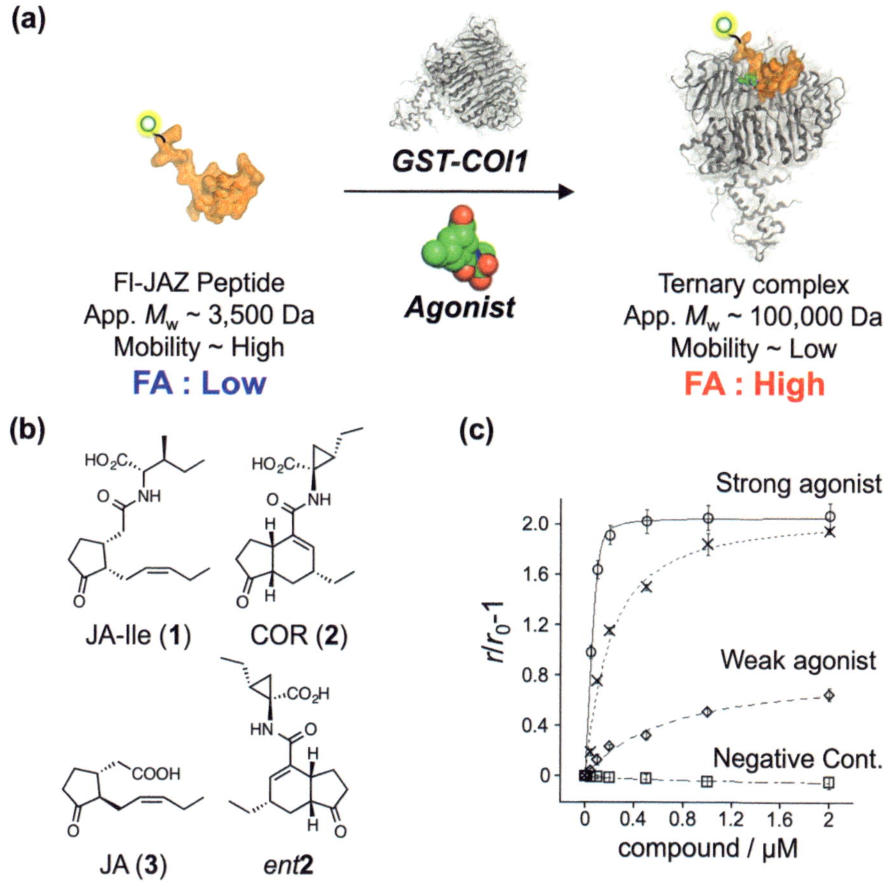

Fig. 4 FA-based in vitro titration assay for COI1-JAZ agonists. (**a**) Schematic of FA change in the absence or presence of the COI1-JAZ agonist based on the significant differences between apparent molecular weights of Fl-JAZ peptide and its ternary complex with GST-COI1 and agonists. (**b**) Chemical structures of JA-Ile (**1**), coronatine (**2**), jasmonic acid (JA, **3**) and the enantiomer of coronatine (*ent***2**). (**c**) Representative FA change of Fl-JAZ peptide with GST-COI1 upon addition of test compounds (circle or cross: strong agonist such as **1** or **2**, diamond: weak agonist, square: negative control such as **3** or *ent***2**). These experiments were performed in triplicate to obtain mean and s.d. (shown as error bars)

Fl-conjugated JAZ peptide (100 nM) in the measurement buffer.

5. Add the ligand dropwise to the solution and then gently mix it with a Pasteur pipette. Then, measure the anisotropy values (r) (*see* **Note 12**).
6. Add the ligand to the solution until the fluorescence anisotropy value is saturated.
7. For each point, calculate an average of the r values from three independent measurements (Fig. 4c).

8. Evaluate apparent K_a values by nonlinear curve-fitting analysis of the anisotropy titration curves and the following equation in KaleidaGraph (Synergy Software, U.S.):

$$K_a = \frac{[HG]}{[H][G]}$$

$$[H]_0 = [H] + [HG]$$

$$[G]_0 = [G] + [HG]$$

$$R\left(\frac{r}{r_0} - 1\right) = R_{max} \cdot \frac{[HG]}{[H][G]} = R_{max} \cdot \frac{[HG]}{[H]_0}$$

$$R = \frac{R_{max}}{2K_a[H]_0}[(K_a[H]_0 + K_a[G]_0 + 1) - \left\{(K_a[H]_0 + K_a[G]_0 + 1)^2 - 4K_a^2[H]_0[G]_0\right\}^{\frac{1}{2}}$$

$[H]$: COI1–JAZ co-receptor concentration, $[G]$" ligand concentration, $[HG]$: ternary complex concentration, R: variation value of fluorescence spectra, and R_{max}: max of the variation value.Then, the K_d values can be calculated taking the reciprocal of the obtained K_a values.

3.5 Fluorescence Anisotropy Titration Experiments of the Antagonists of COI1–JAZ Co-Receptors (Fig. 5)

1. Fluorescence anisotropy is measured by using spectrofluorometer (Subheading 2.3, **step 2**) same as Subheading 3.4, **step 1** and followed by Subheading 3.4, **steps 2–4** as well.
2. To prepare the ternary complex (Fig. 5a), **1** (0–1 μM) is added into the solution containing GST-COI1 (100 nM) and Fl-conjugated JAZ peptide (100 nM), and it is mixed by using Pasteur pipette gently for a few times. And then the r value can be measured.
3. Add the antagonist (e.g., COR-MO (**4**) in Fig. 5b) dropwise to the solution and mix gently with Pasteur pipette. Anisotropy values (r) are then measured.
4. Add the antagonist to the solution until the fluorescence anisotropy value change is saturated. (0–20 μM in case of **4**).
5. Calculate an average of the r values from three independent measurements for each point.
6. Calculate the IC_{50} value with the linear curve-fitting analysis by the plot of $r/r_0 - 1$ versus $\log_{10}$[antagonist].

3.6 Plate-Reader Assay

1. The fluorescence anisotropy change can also be measured using a plate reader equipped with a polarized filter for both excitation and emission fluorescence signals.
2. The Fluorescence Intensity (FL) mode is used in the absence or presence of Fl-JAZ for checking the background signal of the

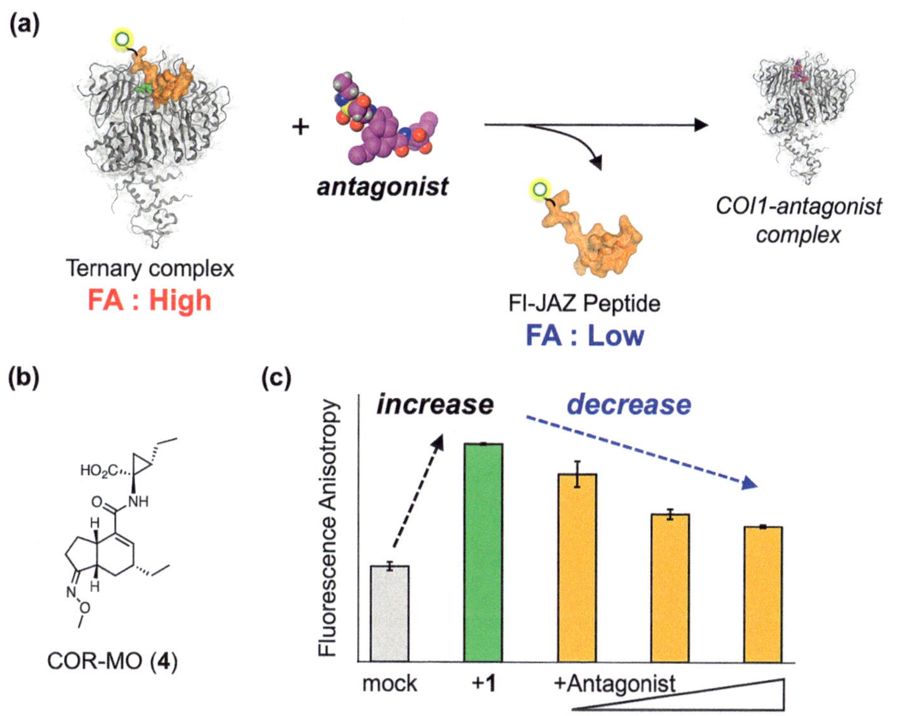

Fig. 5 FA-based in vitro titration assay for COI1/JAZ antagonists. (**a**) Schematic of FA change upon addition of COI1–JAZ antagonist. (**b**) Chemical structure of COR-MO (**4**). (**c**) Representative FA values of Fl-JAZ peptide with GST–COI1 in the absence or presence of **1**, or in the presence of both **1** and antagonist. These experiments were performed in triplicate to obtain mean and s.d. (shown as error bars)

buffer solution and the raw fluorescent intensity of the Fl-JAZ sample. The detection mode is then changed into Fluorescence Polarization (FP) to obtain the fluorescence anisotropy value (r).

3. Add to the 96-well black clear-bottomed plate 100 μL of measurement buffer containing GST–COI1 (50 nM) and Fl-conjugated JAZ peptide (50 nM).
4. Measure an initial anisotropy intensity (r_0) ($\lambda_{ex}/\lambda_{em}$ = 485 nm/516 nm).
5. (Agonist assay only) Add the test compound to the solution and mix by gently mixing with a micropipette. Incubate for 5–30 min at room temperature, then measure the r value (**Note 12**).
6. (Antagonist assay only) Add **1** (0.25 μM) to the solution and mix by gently mixing with a micropipette (it is hard to use Pasteur pipette for microplate). Incubate for 5–30 min at room temperature, then measure the r value The test compound of the antagonist (e.g., COR-MO (**4**), Fig. 5b) is added to the solution in a same well (0–2 μM in case of **4**) and then mixed by

using micropipettes gently for a few times. After incubation for up to 30 min (at least 5 min) at room temperature, the *r* value can be measured.

7. Experiments should be performed in triplicate to obtain mean and s.d. (shown as error bars).

4 Notes

1. Buffers containing some reductants such as 2-mercaptoethanol or reduced glutathione should be prepared freshly for each experiment.
2. Prepared virus solution was stored at 4 °C in dark. P2 virus should be prepared from P1 virus freshly for every 6 months.
3. Cells washed with D-PBS (−) can be stored at −80 °C. Frozen cells were thawed on ice.
4. Purified COI1–GST was dispensed and stored at −80 °C in freezer. Thawed GST–CO1 should not be frozen again.
5. The pull-down assay should be performed for at least three times separately. Pull-down rate can be calculated from the average of each band intensity.
6. Goat anti-fluorescein antibody (0.2 μL, GeneTex, CA, US) can also be used for pull-down assay. In that case, the use of Protein G Mag Sepharose Xtra (25 μL in 50% incubation buffer slurry, GE Healthcare, 28944008) is recommended because of the low cross-reactivity of the antibody.
7. At least in our experiments, SureBeads™ Protein A or G (10 μL in 50% incubation buffer slurry, Bio-Rad, Hercules, CA) can also be used for pull-down assay (Protein A; 1614013, Protein G: 1614023, respectively).
8. Deprotect and elute from the resin methionine- or cysteine-containing peptides (JAZ9 or JAZ13) by stirring with a 90% TFA solution, containing 2% anisole, 5% thioanisole, and 3% 1,2-ethanedithiol, to avoid methionine oxidation.
9. Purified fluorescein-conjugated JAZ peptide was dispensed and stored at −80 °C in freezer. Do not refreeze thawed fluorescein-conjugated JAZ peptide.
10. Store stock solutions of methionine- or cysteine-containing peptides (such as JAZ9 or JAZ13) with tris(2-carboxyethyl) phosphine hydrochloride (TCEP, 1 mM) for avoidance of methionine oxidation.
11. After purification of fluorescein-conjugated JAZ peptide, eluted peptide was assigned with MALDI-TOF MS analysis (4800 plus MALDI TOF/TOF Analyzer, AB Sciex,

Framingham, MA), and purity of the peptide was analyzed using HPLC.

12. For agonist or antagonist assay with spectrofluorometer, after the addition of the ligand, mix the solution gently with a Pasteur pipette to obtain the stable *r* value (*see* Subheadings 3.4 and 3.5).

Acknowledgments

This work was supported by a Grant-in-Aid for Scientific Research (no. 26282207, 17H06407, 17H00885 to MU, and No. 25708026, 18H02101 to Y.T.), JSPS A3 Foresight Program (to MU), JSPS Core-to-core Program 'Asian Chemical Biology Initiative' (to MU), and JST PREST (JPMJPR16Q4 to YT).

References

1. Santner A, Estelle M (2009) Recent advances and emerging trends in plant hormone signalling. Nature 459(7250):1071–1078
2. Chini A, Fonseca S, Fernandez G, Adie B, Chico JM, Lorenzo O, Garcia-Casado G, Lopez-Vidriero I, Lozano FM, Ponce MR, Micol JL, Solano R (2007) The JAZ family of repressors is the missing link in jasmonate signalling. Nature 448(7154):666–671
3. Thines B, Katsir L, Melotto M, Niu Y, Mandaokar A, Liu G, Nomura K, He SY, Howe GA, Browse J (2007) JAZ repressor proteins are targets of the SCF(COI1) complex during jasmonate signalling. Nature 448 (7154):661–665
4. Yan J, Zhang C, Gu M, Bai Z, Zhang W, Qi T, Cheng Z, Peng W, Luo H, Nan F, Wang Z, Xie D (2009) The Arabidopsis CORONATINE INSENSITIVE1 protein is a jasmonate receptor. Plant Cell 21(8):2220–2236
5. Campos ML, Yoshida Y, Major IT, de Oliveira Ferreira D, Weraduwage SM, Froehlich JE, Johnson BF, Kramer DM, Jander G, Sharkey TD, Howe GA (2016) Rewiring of jasmonate and phytochrome B signalling uncouples plant growth-defense tradeoffs. Nat Commun 7:12570
6. Dejonghe W, Russinova E (2017) Plant chemical genetics: from phenotype-based screens to synthetic biology. Plant Physiol 174(1):5–20
7. Toth R, van der Hoorn RAL (2010) Emerging principles in plant chemical genetics. Trends Plant Sci 15(2):81–88
8. Takaoka Y, Iwahashi M, Chini A, Saito H, Ishimaru Y, Egoshi S, Kato N, Tanaka M, Bashir K, Seki M, Solano R, Ueda M (2018) A rationally designed JAZ subtype-selective agonist of jasmonate perception. Nat Commun 9(1):3654
9. Monte I, Hamberg M, Chini A, Gimenez-Ibanez S, Garcia-Casado G, Porzel A, Pazos F, Boter M, Solano R (2014) Rational design of a ligand-based antagonist of jasmonate perception. Nat Chem Biol 10 (8):671–676
10. Matiolli CC, Melotto M (2018) A comprehensive Arabidopsis yeast two-hybrid library for protein-protein interaction studies: a resource to the plant research community. Mol Plant-Microbe Interact 31(9):899–902
11. Yan J, Yao R, Chen L, Li S, Gu M, Nan F, Xie D (2018) Dynamic perception of Jasmonates by the F-box protein COI1. Mol Plant 11 (10):1237–1247
12. Burger M, Mashiguchi K, Lee HJ, Nakano M, Takemoto K, Seto Y, Yamaguchi S, Chory J (2019) Structural basis of Karrikin and non-natural Strigolactone perception in Physcomitrella patens. Cell Rep 26(4):855–865
13. Milroy LG, Grossmann TN, Hennig S, Brunsveld L, Ottmann C (2014) Modulators of protein-protein interactions. Chem Rev 114 (9):4695–4748
14. Takaoka Y, Nagumo K, Azizah IN, Oura S, Iwahashi M, Kato N, Ueda M (2019) A comprehensive in vitro fluorescence anisotropy assay system for screening ligands of the jasmonate COI1-JAZ co-receptor in plants. J Biol Chem 294(13):5074–5081

15. Labbe CM, Laconde G, Kuenemann MA, Villoutreix BO, Sperandio O (2013) iPPI-DB: a manually curated and interactive database of small non-peptide inhibitors of protein-protein interactions. Drug Discov Today 18 (19-20):958–968
16. Sheard LB, Tan X, Mao H, Withers J, Ben-Nissan G, Hinds TR, Kobayashi Y, Hsu FF, Sharon M, Browse J, He SY, Rizo J, Howe GA, Zheng N (2010) Jasmonate perception by inositol-phosphate-potentiated COI1-JAZ co-receptor. Nature 468(7322):400–405
17. Chini A, Gimenez-Ibanez S, Goossens A, Solano R (2016) Redundancy and specificity in jasmonate signalling. Curr Opin Plant Biol 33:147–156
18. Thireault C, Shyu C, Yoshida Y, St Aubin B, Campos ML, Howe GA (2015) Repression of jasmonate signaling by a non-TIFY JAZ protein in Arabidopsis. Plant J 82(4):669–679
19. Mizusawa K, Takaoka Y, Hamachi I (2012) Specific cell surface protein imaging by extended self-assembling fluorescent turn-on nanoprobes. J Am Chem Soc 134 (32):13386–13395

Chapter 12

Use of Fluorescent Reporters to Analyse Dynamic and Spatial Responses to Mechanical Wounding

Antoine Larrieu, Trang Hieu Nguyen, and Antony Champion

Abstract

Mechanical wounding of plant tissues triggers many different responses (Savatin DV, Gramegna G, Modesti V, Front Plant Sci 5:470, 2014). These are primarily mediated by the plant hormone Jasmonic Acid Isoleucine (JA-Ile). Recently, a fluorescent biosensor for JA-Ile showed that sample preparation (i.-e., handling of samples) for fluorescent microscopy very often triggers wound response, even without apparent damage to the seedling, affecting downstream analyses (Larrieu A, Champion A, Legrand J, Nat Commun 6:6043, 2015). In this chapter, we describe how to overcome this technical limitation to monitor any fluorescent reporter or dye in response to wounding, using any type of fluorescent or confocal (inverted or upright, laser scanning or spinning disc) microscopes. Pharmacological or wound treatments can easily be performed and responses monitored over long periods of time. We further describe a simple method to extract and analyse quantitative data from confocal images using the open source software Fiji (Fiji Is Just ImageJ (Schindelin J, Arganda-Carreras I, Frise E, Nat Methods 9:676–682, 2012)) and OpenOffice.

Key words Confocal microscopy, Live imaging, Jasmonic Acid, Arabidopsis, Wound responses, Fluorescent biosensor

1 Introduction

Fluorescent proteins or dyes can be easily monitored, using laser scanning or spinning disc confocal microscopes. Provided microscope settings are carefully selected, samples can be followed for long (>72 h) periods of time without any reduction in fluorescence intensity [1]. These approaches have given new insights into dynamic molecular processes at the cellular scale, particularly those occurring at the second or minute scale [2].

Confocal microscopy has been used on intact living plants, to follow one or several fluorescent reporters at the same time and in response to various treatments [2, 3]. These require to first germinate the seeds on agar plates that are then carefully transferred onto slides or Petri dishes suitable for imaging. While these have no

Antony Champion and Laurent Laplaze (eds.), *Jasmonate in Plant Biology: Methods and Protocols*, Methods in Molecular Biology, vol. 2085, https://doi.org/10.1007/978-1-0716-0142-6_12,

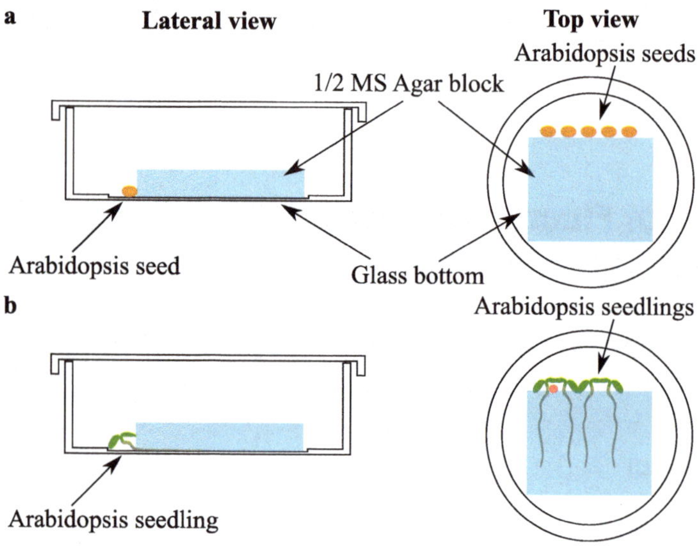

Fig. 1 Setup used for long-term imaging of *Arabidopsis* roots, using confocal microscope in the absence of mechanical stimulations. (**a**) Lateral and top view of the glass-bottom Petri dish after transferring the agar block and sowing seeds against it. (**b**) Lateral and top view of seedlings that have been growing for 4 days after germination in the glass-bottom Petri dishes. Provided the dish has been carefully placed in a suitable growth chamber, most roots will grow in-between the glass bottom and the agar block. The red spot indicates where to apply the treatment if the left-most seedling is used for imaging (Subheading 3.2, **step 1.g**, *see* **Note 9**)

apparent effects on seedlings, it has been shown that such treatments trigger almost immediately and long-lasting JA-mediated wound responses, affecting any downstream experiments on wound responses [4].

In this chapter, we present a method to overcome this limitation. Seeds of a fluorescent reporter of interest are directly sown against a block of agar on Petri dishes fitted with a cover-glass bottom (Fig. 1a). Seedlings of which the roots have grown in between the agar and the glass bottom can be used directly for imaging on a fluorescence (confocal, spinning disc, etc.) microscope without having to manipulate the sample (Fig. 1b). Pharmacological treatments or wounding can be easily performed and fluorescence conveniently followed in growing roots. This protocol is recommended for live imaging of fluorescent reporters in the absence of mechanical perturbations. While this protocol has been used on Arabidopsis thaliana, its principle should be transferable to other plant systems with minor changes. Downstream analyses are described to extract and analyze fluorescence data.

2 Materials

Follow all waste disposal regulations after working with transgenic *A. thaliana* seeds and seedlings.

2.1 A. thaliana Seed Germination and Growth

1. Half-strength Murashige and Skoog (MS) medium: To 2.15 g of MS salts, add approximately 800 mL of double-distilled water (ddH_2O) and adjust the pH to 5.7, using 1 M KOH. To prepare solid medium, top up solution to 1 L and add 10 g of plant agar. To prepare liquid medium, top up solution to 1 L (*see* **Note 1**). Autoclave for 11 min at 121 °C, 20 psi. Autoclaved MS medium can be stored for several months at room temperature. Use a microwave oven to melt solid medium when needed.
2. If interested in following JA responses, seeds for the jasmonates biosensor (Jas9-Venus), as well as the binary plasmid with the p35S-Jas9-Venus construct, are available from the Nottingham Arabidopsis Stock Centre (N2105629, http://arabidopsis.info/StockInfo?NASC_id=2105629 and N2105638, http://arabidopsis.info/StockInfo?NASC_id=2105638). Alternatively, any other fluorescent reporter line can be used.
3. Seed surface sterilization solution: 25% (v/v) sodium hypochlorite, 0.1% (v/v) Triton X-100 in ddH_2O (*see* **Note 2**).
4. Sterile ddH_2O.
5. Sterile cell culture dish with glass bottom, available from numerous manufacturers (e.g., Greiner Bio-One Inc., Germany, 627861; Thermo Fischer, United Kingdom, 150682).
6. Regular sterile square or round Petri dishes.
7. Growth chamber with appropriate lighting and temperature control (we used the following conditions: 24 h daylight, 22 °C, 100 μE light intensity).
8. Micropore surgical tape.

2.2 Image Acquisition and Analysis (See Note 3)

1. Any type of fluorescence or confocal (inverted or upright, laser scanning or spinning disc) microscope fitted with appropriate camera/detectors and ideally with a long distance 10x objective can be used (*see* **Note 4**).
2. Fiji software "Fiji Is Just ImageJ" with ImageJ version 1.52i and Bio-Formats Plugin version 5.9.2 [http://fiji.sc/wiki/index.php/Fiji].
3. Spreadsheet program (e.g., Apache OpenOffice Calc version 4.1.6).

3 Methods

3.1 Seed Germination

1. Surface sterilize seeds expressing one or several fluorescent reporters by soaking in sterilization solution for 8 min. Rinse the seeds three times in sterile ddH_2O. Sterilized seeds can be kept in sterile ddH_2O for a few days at 4 °C.
2. Prepare half-strength MS plates (*see* **Note 5**).
3. Using an appropriate sterile instrument (e.g., spatula, etc.), take a piece of agar approximately 2/3 the size of the glass-bottom Petri dishes (Fig. 1a) and transfer the block directly into one of these (*see* **Note 6**). For each treatment, we recommend preparing four different glass-bottom Petri dishes in case some do not grow well/are contaminated.
4. Using a pipette with a sterile 1 mL tip, sow the sterilized seeds in ddH_2O in a line directly on top of the agar block and against the glass bottom as in (Fig. 1a). Close the plate and secure the lid with Micropore tape. Stratify the seeds on plates at 4 °C for 2 days (not required if seeds have been stored in water for few days at 4 °C).
5. Transfer the plates to a tissue culture room for 4–5 days (*see* **Note 7**).

3.2 Experimental Setup and Image Acquisition

1. Follow these steps to optimise the settings of a confocal microscope:
   (a) If using the JA biosensor Jas9-Venus (*see* Subheading 2.1, **step 2**): Jas9-Venus can be excited with a 514- or a 488-nm laser line. H2B-RFP can be excited using a 561-nm laser line. One must be careful when using the 514 nm line to excite Venus as it will also excite the H2B-RFP fusion protein (up to 30% of the maximum excitation of RFP). If using a 514 laser line, ensure to only collect fluorescence from 520 nm to 550 nm to avoid mixing Venus and RFP fluorescence.
   (b) Make sure to give plenty of time for the lasers to warm-up if they were switched off before starting the experiment.
   (c) Place the glass-bottom Petri dish on the stage and select a region of interest without air bubbles. Make sure to choose the healthiest looking seedlings.
   (d) Set the offset, laser power, detector sensitivity, pinhole, and frame and line averages to obtain the best signal to noise ratio, making sure not to saturate the signal (*see* **Note 8**).
   (e) Choose the smallest time interval between each scan to give a suitable temporal resolution without bleaching the fluorophores. We usually image seedlings every 2 min on a

confocal microscope and every 30 s on a spinning disc, both with limited photo toxicity. Time courses have been carried out for over 36 h using glass-bottom Petri dishes with no effects on Jas9-Venus or H2B-RFP fluorescence. We use a 10× objective in order to maximize the number of nuclei being observed.

(f) Perform a time-course experiment for about 30–60 min without any treatments or touching the seedling.

(g) Using Fiji, perform the analysis described below (Subheading 3.3). If there is less than 20% variation of Venus fluorescence over the time course, the sample can be used for further treatments. To perform a pharmacological treatment, add 100 μL of liquid half-strength MS containing the chemical of interest at the desired concentration next to the seedling (*see* **Note 9**, red spot on Fig. 1b). If control profile varies by more than 20%, *see* **Note 10**.

(h) Scanner and detectors settings used for one experiment are kept unchanged throughout the experiment. When using the same microscope and the same reporter line, it is advised to use similar settings to make sure seedlings have comparable levels of Venus fluorescence at t_0.

3.3 Data Analysis

1. Drag and drop the confocal file to open it with an up-to-date version of Fiji (or go in Plugin > Bio-Formats > Bio-Formats Importer) (*see* **Note 11**).
2. On the Bio-Formats import menu, uncheck all the options except "Split channels" and select "Hyperstack" in the "View stack with" menu, select XYCZT in the "Stack order" menu and select "Default" in the Color mode menu.
3. When imported, choose the channel you want to analyse, select a region of interest (ROI) using one of the selection tool and press CTRL+T to add the ROI to the ROI manager (or go in Edit > Selection > Add to manager). If the analysis is performed on the whole image, draw an ROI encompassing the whole image.
4. Set the measurements (Analyse > Set Measurements) to return the "Integrated Density" and "Min & max grey value." Make sure the options "Limit to threshold" and "Display label" are selected. All the other options are not required to perform the analysis described here, however, depending on your applications, check the relevant options.
5. Increase the upper threshold to the max and set the lower threshold so to discard background pixels (CTRL + SHIFT +T or go in Image > Adjust > Threshold) (*see* **Note 12**). Export the measurements for all the time points by selecting the ROI in the manager, then click on "More" followed by

"Multi Measure." Copy and paste the table into a spreadsheet editor using CTRL+A to select all lines, CTRL+C to copy, and CTRL+V to paste on a worksheet (or go in Edit > Select All in the table window, then Edit > Copy).

6. If analyzing several fluorophores (in the case of the jasmonates biosensor, there is also a reference channel (H2B-RFP) for ratiometric measurements), repeat the procedure using the other channel but with the same ROI. Note that the threshold will most likely be different.
7. If you have a reference fluorophore, normalize the signal of interest (in this case, Jas9-Venus) to the reference (in this case, H2B-RFP) in OpenOffice Calc to obtain a normalized fluorescence.
8. Normalize the values to the first time point to obtain the "Relative Normalized Fluorescence (RNF)."
9. We recommend plotting the RNF of at least five replicates for each treatment to allow buffering for biological variation as well as variations in fluorescence caused by small movements of the sample during the experiment.

4 Notes

1. When performing pharmacological treatments, it is recommended to use liquid medium that has been prepared with the same batch of MS solution used for the agar block to reduce variability. Also, it must be at the same temperature when performing the experiment.
2. Sterilization solution should be prepared fresh each time.
3. Software versions indicated are the ones used for this manuscript, but any previous version should work as only basic functions are required. In doubt, refer to software manuals if using older versions.
4. Additional lighting sources (e.g., LEDs) can be positioned toward the stage of the microscope for long (>2 h) time courses provided the light does not interfere with the detectors.
5. We usually prepare thick plates (30 or 65 mL of molten MS medium per round (9 cm) or square (12 cm) plate, respectively) to make sure the agar block has enough moisture to grow healthy seedlings.
6. Adding a drop of sterile ddH_2O on the glass bottom prior to transfer the agar block avoids the formation of air bubbles that can potentially affect root growth.

7. We usually keep the glass-bottom Petri dishes upright with a slight angle to make sure most roots will grow directly between the bottom of the agar block and the glass bottom (Fig. 1b). Under such conditions, roots usually reach the bottom of the agar block 6 days after transfer to the growth chamber (4–5 days after germination).
8. To make sure background pixels are set to zero, and those corresponding to the fluorophore are not saturating (i.e., the signal is not being clipped at the upper limit of intensity), use a suitable Look Up Table (LUT). Usually, confocal microscope software have specific settings to highlights such pixels.
9. When doing this, touching the root in any way, or moving the seedling around will inevitably lead to a loss of Venus fluorescence in ±30 min. Also, seedlings growing on a contaminated plate or stressed (for instance, when the agar block is drying out) would not have much Venus fluorescence (and usually have high background auto fluorescence). These should be discarded and only healthy seedlings used.
10. If the fluorescence signal changes excessively in the absence of treatments, then there could be too much variation in some of the inherent confocal hardware. First thing is to try and repeat with a different seedling on a different plate. If problems persist, repeat making sure the confocal laser is warmed up, and use a lower laser power setting to avoid photo-bleaching.
11. To carry out the analyses with Fiji, the only required plugin is Bio-Formats (included by default with Fiji). This package allows to open files from many different confocal microscope manufacturers. Make sure to always perform the analyses of all your images with the same version of Fiji.
12. When setting the threshold, the pixels included appear in red. The aim is to select only informative pixels (i.e., remove background ones) for quantification. Do not click on Apply, as it will convert the images to binary. Simply leave the Threshold box open.

Acknowledgments

A.L. is a postdoctoral fellow at the University of Leeds, UK. T.H.-N. is supported by a French Embassy Fellowship.

References

1. Kirchhelle C, Moore I (2017) A simple chamber for long-term confocal imaging of root and hypocotyl development. J Vis Exp (123)
2. Brunoud G, Wells DM, Oliva M et al (2012) A novel sensor to map auxin response and distribution at high spatio-temporal resolution. Nature 482:103–106
3. Federici F, Dupuy L, Laplaze L et al (2012) Integrated genetic and computation methods for in planta cytometry. Nat Methods 9:483–485
4. Larrieu A, Champion A, Legrand J et al (2015) A fluorescent hormone biosensor reveals the dynamics of jasmonate signalling in plants. Nat Commun 6:6043

Chapter 13

Quantitative Jasmonate Profiling Using a High-Throughput UPLC-NanoESI-MS/MS Method

Cornelia Herrfurth and Ivo Feussner

Abstract

Jasmonic acid (JA) and its many derivatives-collectively referred as jasmonates-occur ubiquitously in land plants and regulate a wide range of stress-responses and development. Measuring these signaling compounds is complicated by the large number of jasmonate derivatives and the comparatively low concentration of these metabolites in plant tissues. We, here, present a selective and sensitive method consisting of a two-phase extraction coupled with liquid chromatography, nanoelectrospray ionization, and mass spectrometry to determine jasmonate levels in tissues and fluids of various plant species. The application of stable deuterium-labelled standards in combination with authentic standards allows the absolute quantification of a multitude of jasmonates and, additionally, the semi-quantitative analysis of further metabolites from the jasmonate pathway.

Key words Jasmonic acid, Jasmonates, Liquid chromatography, Nanoelectrospray ionization, Mass spectrometry, Two-phase extraction, Deuterium-labeled standards

1 Introduction

Jasmonates are specialized metabolites formed via the lipoxygenase pathway and originate from plastidic membrane lipids, preferentially from α-linolenic acid [1]. The central compound of this metabolic pathway is jasmonic acid (JA) that can be converted to a multitude of metabolites through methylation, glycosylation, hydroxylation, sulfation, *O*-glycosylation, or esterification with amino acids [2]. This includes the isoleucine conjugate of JA (JA-Ile) currently known as the unique JA derivative directly responsible for JA signaling. At least 12 metabolic pathways have been described so far to convert JA or derivatives from JA, leading to active, inactive, and partially active compounds [3]. Jasmonates exist ubiquitously in land plants and act as important regulators in the plants' response to biotic and abiotic stresses as well as in development like root growth, seed germination, stamen development, or senescence.

Antony Champion and Laurent Laplaze (eds.), *Jasmonate in Plant Biology: Methods and Protocols*, Methods in Molecular Biology, vol. 2085, https://doi.org/10.1007/978-1-0716-0142-6_13, © Springer Science+Business Media, LLC, part of Springer Nature 2020

Jasmonates are typically present only at trace levels (pmol–nmol/g fresh weight [f.w.]) in plant tissues, and this makes a qualitative and quantitative analysis of these compounds challenging. In the beginning, JA was detected by radio immunoassays and enzyme-linked immunosorbent assays [4]. These methods are sensitive, but cross-reactivity of the required antibodies can result in reduced specificity and accuracy. In recent years, chromatographic separation combined with mass spectrometry (MS) has become the most powerful technique for the identification and characterization of the very low abundant jasmonates, with low interference with other matrix compounds. Gas chromatography (GC)–MS has been widely used to determine the levels of endogenous jasmonates and thereafter their involvement in plant stress signaling [5]. However, chemical derivatization of hydrophilic groups is required to increase the volatility of the analytes, and the high temperature in the GC injector and columns can also degrade thermally unstable compounds [6]. In comparison to GC–MS, liquid chromatography (LC)–MS has the advantage that samples can usually be analyzed without prior time-consuming derivatization procedures as required for GC–MS approaches. Therefore, reversed phase–based LC separation interfaced with electrospray ionization (ESI) for tandem mass spectrometry (MS/MS) has become more and more popular for the unambiguous detection more recently. Many publications have described sensitive and selective jasmonate profiling methods, using the highly selective multiple reaction monitoring (MRM) mode (exemplary beside many others: [7–9]). The sensitivity and reproducibility of a LC–MS/MS method can be furthermore increased using a chip-based nanoESI source. With this device, the contamination of the MS system is reduced to a very low limit, therefore, requiring far less frequent MS cleaning.

The methodology reported herein was developed for the rapid and sensitive quantification of jasmonates (Table 1) from a minimal amount of plant material, and it additionally covers the targeted analysis of the biosynthetic precursors of JA, namely 12-oxo-phytodienoic acid (oPDA), its 16-carbon homolog 10-oxo-dinor-phytodienoic acid (dinor-oPDA), as well as their β-oxidation intermediates toward JA formation. The workflow includes a two-phase methyl-*tert*-butyl ether (MTBE):methanol:water-based extraction procedure [10] as modified for the metabolite fingerprinting approach [11], but using the hydrophobic phase only for jasmonate analysis, here. For the quantitative analysis, three deuterium-labeled jasmonate compounds are added before extraction to the homogenized plant material. These isotope-labeled internal standards have a very similar chemistry as the target jasmonates, leading to almost identical physicochemical properties and both chromatographic and MS behavior. The addition of the internal standards provides the correction of the jasmonate contents for effects related to the plant matrix and for losses during sample

Table 1
Abbreviation, name, and structure of jasmonate compounds and internal standards used for quantification (absolute) and semi-quantification (relative), respectively including their commercial availability

Abbreviation	Name	Structure	Internal standard	Quantification	Catalogue number[b]
oPDA	12-oxo phytodienoic acid	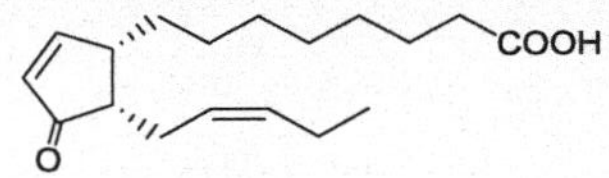	D_5-OPDA	Absolute	0146191
oPDA-Ile	12-oxo phytodienoyl isoleucine	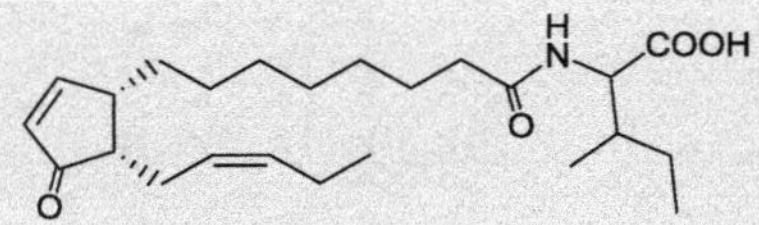	D_5-OPDA	Relative	[a]
dinor-oPDA	10-oxo dinor-phytodienoic acid	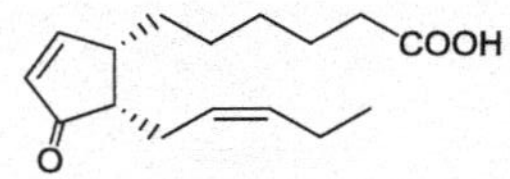	D_5-OPDA	Absolute	0146211
OPC8	3-oxo-2-(2-(*Z*)-pentenyl) cyclopentane-1-octanoic acid	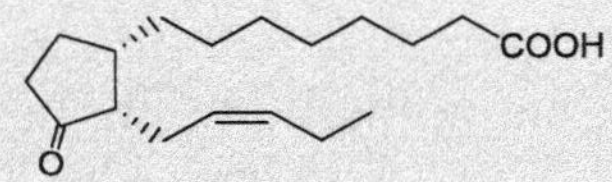	D_5-JA	Absolute	0146201
OPC6	3-oxo-2-(2-(*Z*)-pentenyl) cyclopentane-1-hexanoic acid	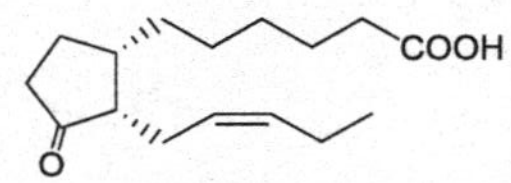	D_5-JA	Absolute	0146181
OPC4	3-oxo-2-(2-(*Z*)-pentenyl) cyclopentane-1-butyric acid	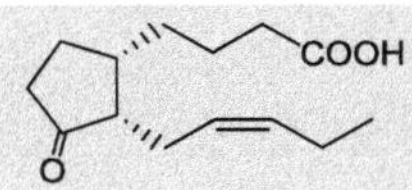	D_5-JA	Absolute	0146221
JA	Jasmonic acid	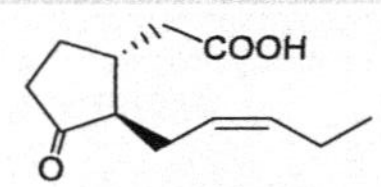	D_5-JA	Absolute	0142821
11OH-JA	11-hydroxy jasmonic acid	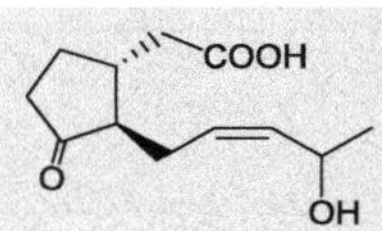	D_5-JA	Absolute	[a]

(continued)

Table 1
(continued)

Abbreviation	Name	Structure	Internal standard	Quantification	Catalogue number[b]
12OH-JA	12-hydroxy jasmonic acid (tuberonic acid)	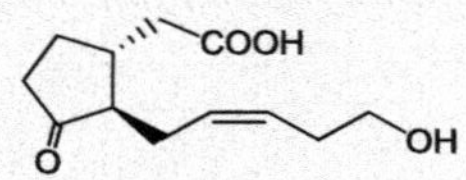	D_5-JA	Absolute	0142871
ddh-JA	4,5-didehydro jasmonic acid	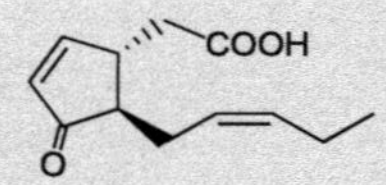	D_5-JA	Absolute	[a]
12HSO4-JA	12-hydroxy jasmonoyl sulfate	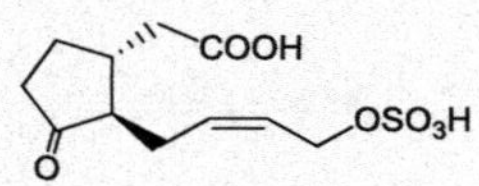	D_5-JA	Relative	[a]
12O-Glc-JA	12-*O*-glycosyl jasmonic acid	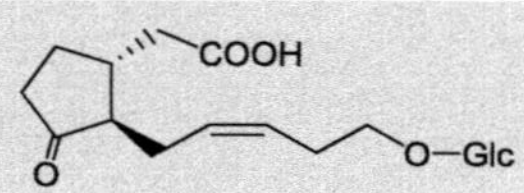	D_5-JA	Relative	[a]
12OH-JA-Ile	12-hydroxy jasmonoyl isoleucine	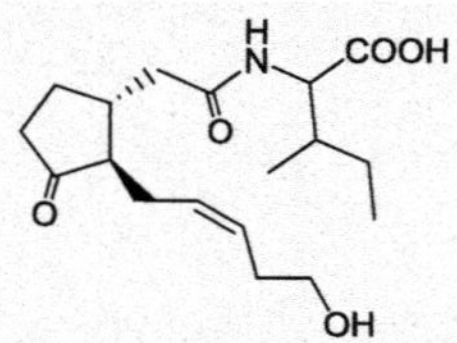	D_5-JA	Relative	[a]
12OH-JA-Val	12-hydroxy jasmonoyl valine	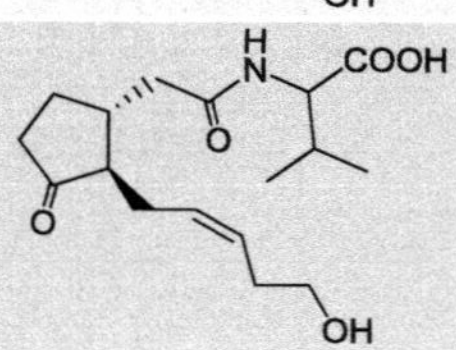	D_5-JA	Relative	[a]

12OH-JA-Phe	12-hydroxy jasmonoyl phenylalanine	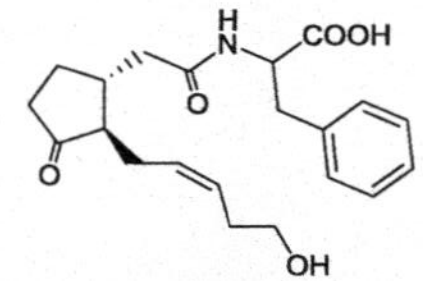	D_5-JA	Relative	[a]
12COOH-JA	12-carboxy jasmonic acid	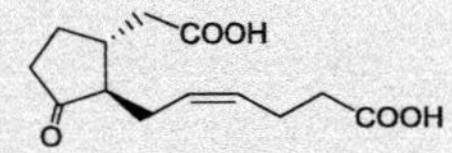	D_5-JA	Relative	[a]
12COOH-JA-Ile	12-carboxy jasmonoyl isoleucine	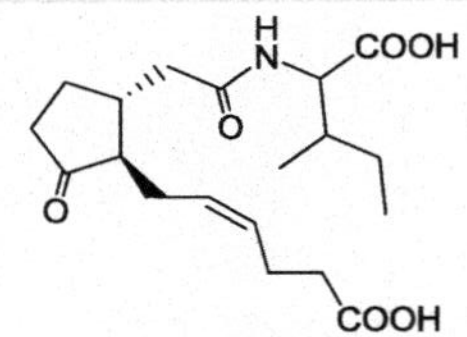	D_5-JA	Relative	[a]
JA-Ile	Jasmonoyl isoleucine	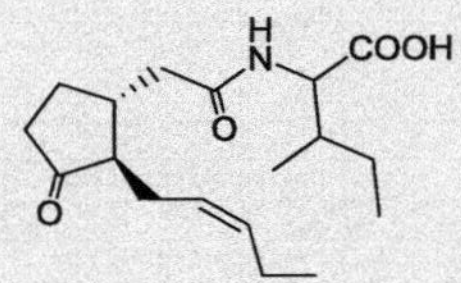	D_3-JA-Leu	Absolute	0146861
JA-Leu	Jasmonoyl leucine	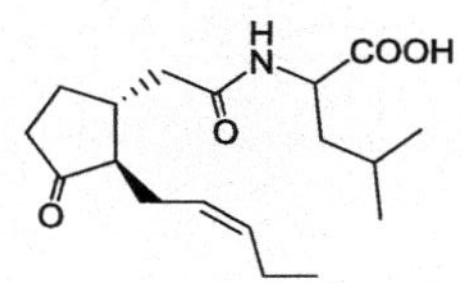	D_3-JA-Leu	Absolute	0146871
JA-Phe	Jasmonoyl phenylalanine	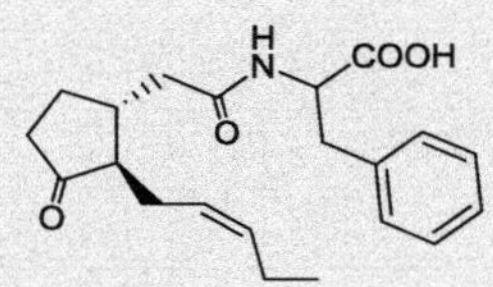	D_3-JA-Leu	Absolute	0146881

(continued)

Table 1 (continued)

Abbreviation	Name	Structure	Internal standard	Quantification	Catalogue number[b]
JA-Val	Jasmonoyl valine	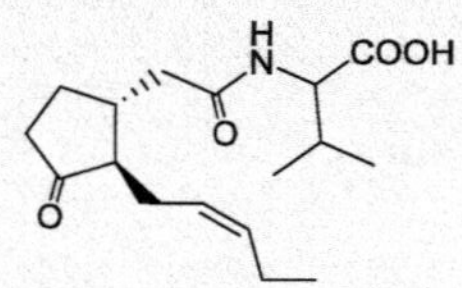	D_3-JA-Leu	Absolute	0146891

[a]No commercial source available so far

[b]Catalogue number from OlChemIm (Olomouc, Czech Republic)

preparation. Quantification was carried out using a calibration curve of intensity (m/z) ratios of (unlabeled)/(deuterium-labeled) versus molar amounts of unlabeled (0.3–1000 pmol) compounds. Each analyte is characterized by the most intense precursor-fragment ion combination and a characteristic retention time. The second most intense transition is used for compound verification, and accurate mass measurements of the precursor ion are done for confirming the compound identity if the compound is identified in another plant species or plant organ for the first time.

2 Materials

2.1 Samples

1. Flash-frozen plant tissue material (leaves, flowers, roots, seedlings, stems), keep at −80 °C until grinding and extraction begins.
2. Xylem sap from plants, keep at −80 °C until the extraction begins.

2.2 Chemicals and Standards

Analytical standards were mostly purchased from OlChemIm (Olomouc, Czech Republic, Tables 1 and 2). Selected jasmonate compounds were provided by Otto Miersch (Halle/Saale, Germany), Robert Kramell (Halle/Saale, Germany), and Mats Hamberg (Stockholm, Sweden). Table 3 lists the relevant publications, describing the LC–MS/MS detection parameters in negative ESI mode for all compounds.

2.3 Solvents and Solutions

All solvents used (methanol, acetonitrile) are LC–MS grade quality unless indicated otherwise. Water is always freshly generated by the Arium pro VF TOC ultrapure water system (Sartorius, Goettingen, Germany). Methyl-*tert*-butylether (MTBE, Sigma-Aldrich, Germany) and 2-propanol are HPLC grade.

1. Solvent system for UPLC analyses with the ACQUITY HSS T3 column (Waters Corporation, Milford, MA, USA): solvent A (water containing 0.3 mmol/L NH_4HCOO, adjusted to pH 3.5 with formic acid), solvent B (acetonitrile/water, 90:10 [*v/v*] containing 0.3 mmol/L NH_4HCOO, adjusted to pH 3.5 with formic acid), solvent for nanoESI stabilization (2-propanol/acetonitrile/water, 70:20:10 [*v/v/v*], containing 0.3 mmol/L NH_4HCOO, adjusted to pH 3.5 with formic acid).
2. Tuning mixture for 4000 QTRAP: Standards chemical kit with low/high concentration PPGs (AB Sciex, Framingham, MA, USA).
3. Internal standard mixture (Table 2) and jasmonate compounds for calibration samples (Table 4).

Table 2
Internal standard mixture of three deuterium-labelled compounds for quantification of jasmonates

Internal standard	Catalogue number[a]	μl/ml stock
D_5-oPDA	0366191	600
D_3-JA-Leu	0366271	200
D_6-JA	0362821	200

Each standard is dissolved individually as a stock solution of 1 ng/μl in ethanol and then combined in the volumes indicated to produce the internal standard mixture
[a]Catalogue number from OlChemIm (Olomouc, Czech Republic)

2.4 LC–MS System

1. For chromatographic separation: ACQUITY UPLC® system (Waters Corp., Milford, MA, USA) equipped with an ACQUITY UPLC® HSS T3 column (100 mm × 1 mm, 1.8 μm; Waters Corp., Milford, MA, USA).
2. For nanoESI: chip ion source TriVersa Nanomate® (Advion, Inc., Ithaca, NY, USA) and Agilent 1100 HPLC isocratic pump (Agilent, Waldbronn, Germany) with mixing tee valve and equipped with nanoESI chip with 5 μm internal diameter nozzles.
3. For mass-spectrometric detection: AB Sciex 4000 QTRAP® tandem mass spectrometer (AB Sciex, Framingham, MA, USA).

2.5 Software

1. Data acquisition: Analyst 1.6.2 (AB Sciex, Framingham, MA, USA).
2. nanoESI controlling: ChipSoft 8.3.1 (Advion, Inc., Ithaca, NY, USA).
3. Data analysis: MultiQuant 3.0.2 (AB Sciex, Framingham, MA, USA).
4. Data processing and statistics: Excel 2016 (Microsoft Corporation, Redmond, WA, USA) and RStudio (RStudio, Inc., Boston, MA, USA).

2.6 Other Equipment

1. Eppendorf tubes (1.5 ml and 2 ml).
2. Kimble extraction tubes (Kimax-51, 13 × 100 mm) with Teflon-coated screw caps (Gerresheimer Glass Inc., Vineland, NY, USA).
3. Adjustable pipettes, including suitable tips and Pasteur pipettes.
4. Chemical resistant tips for MTBE (Safe Seal Tips Premium, Biozym, Oldendorf, Germany).
5. Chemical resistant dispenser for MTBE (Dispensette Organic, Brand GmbH, Wertheim, Germany).

Table 3
Mass transitions and optimized MS parameters for detection of jasmonate compounds by mass spectrometry

Analyte	RT [min]	Q1 [m/z]	Q3 (Quantifier) [m/z]	DP [V]	EP [V]	CE [V]	CXP [V]	Q3 (Qualifier) [m/z]	DP [V]	EP [V]	CE [V]	CXP [V]	Reference
oPDA	6.6	291	165	−50	−5	−26	−2	247	−50	−5	−26	−2	[12]
oPDA-Ile	6.3	404	130	−50	−5	−26	−2	165	−50	−5	−26	−2	[13]
dinor-oPDA	6.4	263	165	−40	−5	−20	−2	219	−40	−5	−26	−2	[12]
OPC8	6.7	293	225	−80	−4.5	−30	−2	249	−80	−4.5	−30	−2	[12]
OPC6	6.5	265	221	−120	−6	−25	0	165	−120	−6	−30	0	[b]
OPC4	5.7	237	165	−110	−6	−24	−2	125	−110	−6	−26	−2	[14]
JA	4.1	209	59	−30	−4.5	−24	−6	165	−30	−4.5	−20	-6	[15]
11OH-JA 12OH-JA	2.5 2.5	225 225	59 59	−35 −35	−9 −9	−28 −28	−6 −6	165 97	−35 −35	−9 −9	−16 −20	−2 −2	[b]
ddh-JA	2.1	207	163	−30	−4.5	−16	−6	93	−30	−4.5	−22	−6	[b]
12HSO4-JA	2.2	305	97	−30	−4	−32	0	225	−35	−4	−26	0	[16]
12O-Glc-JA	2.2	387	59	−85	−9	−52	−6	207	−85	−9	−30	−2	[b]
12OH-JA-Ile	3.2	338	130	−45	−10	−30	−6	308	−45	−10	−30	−6	[17]
12OH-JA-Val[a]	2.8	324	116	−45	−10	−30	−6						[18]
12OH-JA-Phe[a]	3.3	372	164	−45	10	−30	−6						[18]
12COOH-JA	2.5	239	59	−35	−9	−28	−6	195	−35	−9	−28	−6	[19]

(continued)

Table 3
(continued)

Analyte	RT [min]	Q1 [m/z]	Q3 (Quantifier) [m/z]	DP [V]	EP [V]	CE [V]	CXP [V]	Q3 (Qualifier) [m/z]	DP [V]	EP [V]	CE [V]	CXP [V]	Reference
12COOH-JA-Ile	3.2	352	130	−45	−10	−30	−6	308	−45	−10	−30	−6	[17]
JA-Ile	5.2	322	130	−45	−5	−28	−2	172	−50	−5	−22	−2	[20]
JA-Leu	5.4	322	130	−45	−5	−28	−2	172	−50	−5	−22	−2	[21]
JA-Phe	5.2	356	164	−45	−6	−28	−2	97	−40	−6	−46	−2	[22]
JA-Val	4.3	308	116	−45	−5	−28	−2	158	−45	−5	−22	−2	[23]

RT is the retention time, Q1 and Q3 are the parent and product ion (quantifier for quantification and qualifier for compound verification), and CP, EP, CE and CXP indicate the declustering potential, entrance potential, collision energy and cell exit potential for each jasmonate compound

[a]No standard substances available for determining a qualifier mass

[b]unpublished data of the authors

Table 4
Final concentrations of the jasmonate compounds within the calibration sample series and the indicated volumes are used to create the sample series

Final concentration [ng/100 μl]	Volume [μl]
0.1	1.3[a]
0.5	6.5[a]
2	2.6
10	13
20	26
100	130

[a]Use 1:10-diluted calibration stock solution

6. Glass micro vials for analysis (12 mm, 250 μl, Macherey-Nagel GmbH, Dueren, Germany) fixed by a spring in HPLC glass vials (1.5 ml, N9, Macherey-Nagel GmbH, Dueren, Germany).
7. Eppendorf Microcentrifuge 5415D (Eppendorf, Hamburg, Germany).
8. Eppendorf Centrifuge 5810R (Eppendorf, Hamburg, Germany).
9. Nitrogen evaporator (Organomation Associates, Inc., Berlin, MA, USA).
10. Mixer Ball Mill MM200 with stainless steel grinding jars or PTEE-jars (Retsch GmbH, Haan, Germany).
11. Porcelain mortar and pistil.
12. Hood.
13. Argon.

3 Methods

3.1 Harvesting Plant Material

1. Keep the harvesting procedure of plant material as short as possible and always in the same time range. It should not exceed 30 s since JA starts to accumulate after wounding very rapidly [24].
2. Shock freeze the harvested material immediately in liquid nitrogen (*see* **Note 1**).

3.2 Homogenization of Plant Material

1. Grind the plant material by using a porcelain mortar and pistil or homogenize it by using the Mixer Ball Mill MM200 (*see* **Note 2**).

2. Use stainless steel grinding jars (for large sample amounts) or PTEE jars for Eppendorf cups (for small sample amounts) with the corresponding size of stainless steel balls for mill homogenization.
3. Time and repetitions of the homogenization cycles depend on the amount and rigidity of the biological material (*see* **Note 3**).
4. Ensure that the biological material stays always completely frozen under liquid nitrogen.

3.3 Extraction of Plant Material

The two-phase extraction method with MTBE, methanol, and water as described by Matyash et al. [10] was modified for the extraction of plant material as follows:

1. Prior to use, Kimble glass tubes are kept on ice. All solvents are kept at room temperature.
2. Weigh 100 mg of homogenized and deeply frozen material into Kimble glass tubes. Immediately add 0.75 ml of methanol and vortex. Ensure that the biological material stays frozen during weighing.
3. Add 2.5 ml of MTBE (*see* **Note 4**). Ensure that all the materials are washed down from the side of the tube. Overlay with argon. All following steps are performed at room temperature, unless indicated otherwise.
4. Vortex the sample strongly.
5. Add 50 μl of the internal standard solution (Table 2).
6. Shake for 1 h at 6 °C in the dark.
7. Add 0.6 ml of water and vortex.
8. Incubate the samples for 10 min at room temperature.
9. Centrifuge the samples for 15 min at 800 × *g*.
10. Collect the upper (nonpolar) phase with a Pasteur glass pipette and transfer it into a new Kimble glass tube.
11. For re-extraction, add 0.7 ml of methanol/water (3:2.5 (v/v)) to the lower phase and vortex.
12. Add 1.3 ml of MTBE to the lower phase and vortex.
13. Incubate for 10 min and centrifuge for 15 min at 800 × *g*.
14. Collect the upper phase with a Pasteur glass pipette and combine both upper phases.
15. Dry the combined upper extraction phases under a stream of nitrogen.
16. Resolve the samples in 500 μl of methanol and vortex. Ensure that all the sample materials have been removed from the glass border.

17. Transfer the mixture into a 1.5 ml Eppendorf tube, centrifuge (2 min, 16,000 × *g*) and dry carefully under a stream of nitrogen.
18. Resolve the sample in 20 μl of solvent B and vortex. Add 80 μl of solvent A, vortex and centrifuge (10 min, 16,000 × *g*).
19. Transfer 80 μl of the sample into a glass micro-vial.
20. Cover the samples with argon and use immediately for phytohormone analysis (*see* **Note 5**).

To avoid autoxidation, immediately cover the samples with argon after each extraction step, particularly after the last solvation step.

3.4 Extraction of Phytohormones from Aqueous Solutions of Plants

The two-phase extraction method with MTBE, methanol, and water as described by Matyash et al. [10] was modified for the extraction of plant fluids as follows:

1. Thaw the deeply frozen fluids on ice. Vortex the samples after complete thawing.
2. Give 150 μl of methanol to a glass vial; add 200 μl of the sample and vortex. All following steps are performed at room temperature, unless indicated otherwise.
3. Add 500 μl MTBE (*see* **Note 4**) and vortex.
4. Add 30 μl of the internal standard solution (Table 2).
5. Shake for 1 h in the dark.
6. Add 120 μl of water (to induce phase separation) and vortex.
7. Incubate the samples for 10 min.
8. Centrifuge the samples for 15 min at 800 × *g*.
9. Collect the upper (nonpolar) phase with a Pasteur glass pipette and transfer it into a new Kimble glass tube.
10. Dry the phase carefully under a stream of nitrogen.
11. Resolve the sample in 12 μl of solvent B and vortex. Add 48 μl of solvent A, vortex, and centrifuge (10 min, 16,000 × *g*).
12. Transfer 50 μl of the sample into a glass micro-vial.
13. Cover the sample with argon and use immediately for phytohormone analysis (*see* **Note 5**).

3.5 UPLC Parameters

The performance of the UPLC should be controlled regularly by inspecting the back pressure of the system and the retention time stability with one of the calibration samples (Table 4).

1. Set the temperature of the sample manager to 18 °C and the column oven temperature to 40 °C.
2. Set the flow rate to 0.16 ml/min and the injection volume to 3 μl.

Table 5
Solvent gradient for UPLC separation (*see* Note 6) of jasmonates prior to detection by mass spectrometry

Time (min)	Solvent A (%)	Solvent B (%)
0	90	10
0.5	90	10
2.5	60	40
4.5	60	40
5.5	5	95
8	5	95
8.5	90	10
11.5	90	10

3. Use acetonitrile as strong and acetonitrile/water (1:9 (v/v)) as weak wash solution.
4. Use the gradient shown in Table 5 as mobile phase for the chromatographic separation [25].

3.6 Nanoelectrospray Ionization Parameters

The performance of the nanoESI device has to be controlled regularly by visually inspecting the surface of the chip and calibration of the LC coupler using the ChipSoft routine. During the analysis, keep the nanoelectrospray current constantly under control.

1. Set the flow rate of solvent for nanoESI stabilization to 0.07 ml/min at the same time, when the UPLC flow is started.
2. Set the ionization voltage to −1.55 kV in negative nanoESI mode.

3.7 Mass Spectrometry Parameters

The performance of the 4000 QTRAP® tandem mass spectrometer has to be controlled regularly. Inspect the mass sensitivity by using one of the calibration samples (Table 4) before running sample sets. The mass spectrometer has to be cleaned once per year, following the instruction of the manufacturer, and, afterwards, the mass accuracy and resolution has to be tuned using the tuning mixture.

1. Operate the 4000 QTRAP® tandem mass spectrometer in the scheduled MRM mode in the negative mode.
2. Use the mass transitions shown in Table 3 (Quantifier) and Table 6 for the MS detection with an MRM detection window of 72 s and a target scan time of 1.2 s.
3. Adjust the resolution of the mass analyzers to 0.7 amu full width at half-height.

Table 6
Mass transitions and optimized MS parameters for detection of deuterium-labelled internal standards by mass spectrometry

Internal standard	RT [min]	Q1 [*m/z*]	Q3 [*m/z*]	DP [V]	EP [V]	CE [V]	CXP [V]
D_5-oPDA	6.6	296	170	−65	−4	−28	−2
D_6-JA	4.1	215	59	−35	−8.5	−24	−6
D_3-JA-Leu	5.3	325	133	−65	−4	−30	−2

RT is the retention time; Q1 and Q3 are the parent and product ion; and CP, EP, CE, and CXP indicate the declustering potential, entrance potential, collision energy, and cell exit potential for each internal standard

4. Set the ion source temperature to 40 °C and the curtain gas at 10 (given in arbitrary units).

For identification of precursor ions and fragment ions:

1. Use the Q1 MS mode or the product ion mode and vary the declustering potential and collision energy depending on the requirements of the analyte ion.

3.8 Standard Calibration

For absolute quantification and semi-quantification, the UPLC–nanoESI-MS/MS system has to be calibrated regularly every 6 months. The calibration is mandatory after performing the cleaning and tuning routine.

1. Combine 50 μl of 10 ng/μl solutions from each jasmonate compound available with defined concentration (*see* Table 7) to produce a calibration stock solution.
2. Generate 50 μl of a 1:10-diluted calibration stock solution with ethanol.
3. Build up a calibration sample series with increasing amounts of each jasmonate compound in triplicates corresponding to Table 4, using the indicated volumes of the 1:10-diluted calibration stock solution or calibration stock solution, respectively.
4. Transfer the mixtures into 1.5 ml Eppendorf tubes, add 50 μl of the internal standard solution (Table 2), and dry carefully under a stream of nitrogen.
5. Resolve the samples in 20 μl of solvent B and vortex. Add 80 μl solvent A, vortex and centrifuge (10 min, 16,000 × *g*).
6. Transfer 80 μl of the sample into a glass micro-vial.
7. Cover the samples with argon and use immediately for phytohormone analysis (*see* **Note 5**).
8. Perform peak integration with MultiQuant (*see* Subheading 3.9) and plot the ratios of the peak area of (unlabeled)/(deuterium-labeled) versus molar amounts of unlabeled compounds

Table 7
Jasmonate compounds available as 10 ng/μl solutions in ethanol and used for the calibration stock solution

Analyte
oPDA
dinor-oPDA
OPC8
OPC6
OPC4
JA
11/12OH-JA
ddh-JA
JA-Ile
JA-Leu
JA-Phe
JA-Val

into a *xy* chart. Calculate the steepness of the trendline with Excel and use it as calibration value.

3.9 Data Analysis and Processing

For data analysis, perform peak integration with MultiQuant.

1. Set Gaussian smooth width to 1 point.
2. Copy the entire table to Excel.

For data processing, calculate the absolute amounts of the phytohormone species, using the calibration values of the respective standard compounds with Excel. If the standard compound is not available in a defined concentration or if the standard compound is not present at all, calculate relative amounts, using the calibration standard curve of the structurally similar metabolite (Table 1).

3.10 Verification of Newly Identified Phytohormone Species in Distinct Genetic or Plant Species–Specific Backgrounds

The chemical identity of putative phytohormone species has to be confirmed by additional fragmentation ions beside the already used precursor–fragment ion combination for quantification. Additional MRMs (Table 3, Qualifier) provide important information to confirm the identity of putative phytohormone species in distinct genetic or plant species–specific backgrounds.

3.11 Confirmation of Phytohormone Identity by Accurate Mass Measurements

In addition to further fragment ions, the accurate mass of the precursor ion and the fragment ions should be confirmed by high-resolution MS.

4 Notes

1. Wear glasses and cold-protective gloves when handling liquid nitrogen.
2. Cool down all equipment (mortar, pistil, mill jars, cups, tubes, spatula) with liquid nitrogen to avoid thawing and clogging of the sample on the equipment.
3. *Arabidopsis* rosettes are usually completely homogenized to fine powder by using the Mixer Ball Mill MM200 for 1 min at 30 vibrations s^{-1}.
4. For working with MTBE, use pipette tips or dispensers, which are chemical resistant to organic solvents.
5. After preparation of the sample for the UPLC–nanoESI-MS/MS analysis, it has to be analyzed within 2 weeks due to degradation of the D_5-OPDA standard in the water-containing solvent mixture.
6. The isocratic step between 2.5 min and 4.5 min within the UPLC gradient was introduced to sufficiently separate JA-Ile and JA-Leu.

Acknowledgments

We are very grateful to Tim Iven for improving the phytohormone analysis platform and Krzysztof Zienkiewicz for testing the robustness of the method. We thank Sabine Freitag and Pia Meyer for excellent assistance. This research has been funded by the DFG (ZUK 45/2010; INST 186/822-1).

References

1. Wasternack C, Song S (2017) Jasmonates: biosynthesis, metabolism, and signaling by proteins activating and repressing transcription. J Exp Bot 68(6):1303–1321. https://doi.org/10.1093/jxb/erw443
2. Wasternack C, Hause B (2013) Jasmonates: biosynthesis, perception, signal transduction and action in plant stress response, growth and development. An update to the 2007 review in annals of botany. Ann Bot 111(6):1021–1058. https://doi.org/10.1093/aob/mct067
3. Wastermack C (2018) Jasmonates - news on occurrence, biosynthesis, metabolism and action of an ancient group of signaling compounds. Int J Mol Sci 19(9):2539. https://doi.org/10.3390/ijms19092539
4. Kramell R, Atzorn R, Schneider G, Miersch O, Brückner C, Schmidt J, Sembdner G, Parthier B (1995) Occurrence and identification of jasmonic acid and its amino acid conjugates

induced by osmotic stress in barley leaf tissue. J Plant Growth Regul 14(1):29. https://doi.org/10.1007/bf00212643
5. Mueller MJ, Brodschelm W (1994) Quantification of jasmonic acid by capillary gas chromatography-negative chemical ionization-mass spectrometry. Anal Biochem 218(2):425–435. https://doi.org/10.1006/abio.1994.1202
6. Birkemeyer C, Kolasa A, Kopka J (2003) Comprehensive chemical derivatization for gas chromatography-mass spectrometry-based multi-targeted profiling of the major phytohormones. J Chromatogr A 993(1–2):89–102. https://doi.org/10.1016/S0021-9673(03)00356-X
7. Floková K, Tarkowská D, Miersch O, Strnad M, Wasternack C, Novák O (2014) UHPLC–MS/MS based target profiling of stress-induced phytohormones. Phytochemistry 105:147–157. https://doi.org/10.1016/j.phytochem.2014.05.015
8. Trapp MA, De Souza GD, Rodrigues-Filho E, Boland W, Mithöfer A (2014) Validated method for phytohormone quantification in plants. Front Plant Sci 5:417. https://doi.org/10.3389/fpls.2014.00417
9. Pan X, Welti R, Wang X (2010) Quantitative analysis of major plant hormones in crude plant extracts by high-performance liquid chromatography-mass spectrometry. Nat Prot 5(6):986–992. https://doi.org/10.1038/nprot.2010.37
10. Matyash V, Liebisch G, Kurzchalia TV, Shevchenko A, Schwudke D (2008) Lipid extraction by methyl-tert-butyl ether for high-throughput lipidomics. J Lipid Res 49(5):1137–1146. https://doi.org/10.1194/jlr.D700041-JLR200
11. Feussner K, Feussner I (2019) Comprehensive LC-MS-based metabolite fingerpriting approach for plant and fungal-derived samples. Methods Mol Biol 1978:167–185. https://doi.org/10.1007/978-1-4939-9236-2_11
12. Bao J, Gao X, Jones AD (2014) Unusual negative charge-directed fragmentation: collision-induced dissociation of cyclopentenone oxylipins in negative ion mode. Rapid Commun Mass Spectrom 28(5):457–464. https://doi.org/10.1002/rcm.6803
13. Floková K, Feussner K, Herrfurth C, Miersch O, Mik V, Tarkowská D, Strnad M, Feussner I, Wasternack C, Novák O (2016) A previously undescribed jasmonate compound in flowering *Arabidopsis thaliana* – The identification of cis-(+)-OPDA-Ile. Phytochemistry 122(2):230–237. https://doi.org/10.1016/j.phytochem.2015.11.012
14. Glauser G, Wolfender J-L (2013) A non-targeted approach for extended liquid chromatography-mass spectrometry profiling of free and esterified jasmonates after wounding. In: Goossens A, Pauwels L (eds) Jasmonate signaling: methods and protocols. Humana Press, Totowa, NJ, pp 123–134. https://doi.org/10.1007/978-1-62703-414-2_10
15. Durgbanshi A, Arbona V, Pozo O, Miersch O, Sancho JV, Gomez-Cadenas A (2005) Simultaneous determination of multiple phytohormones in plant extracts by liquid chromatography-electrospray tandem mass spectrometry. J Agric Food Chem 53(22):8437–8442. https://doi.org/10.1021/jf050884b
16. Gidda SK, Miersch O, Levitin A, Schmidt J, Wasternack C, Varin L (2003) Biochemical and molecular characterization of a hydroxyjasmonate sulfotransferase from *Arabidopsis thaliana*. J Biol Chem 278(20):17895–17900. https://doi.org/10.1074/jbc.M211943200
17. Glauser G, Grata E, Dubugnon L, Rudaz S, Farmer EE, Wolfender JL (2008) Spatial and temporal dynamics of jasmonate synthesis and accumulation in Arabidopsis in response to wounding. J Biol Chem 283(24):16400–16407. https://doi.org/10.1074/jbc.M801760200
18. Kitaoka N, Kawaide H, Amano N, Matsubara T, Nabeta K, Takahashi K, Matsuura H (2014) CYP94B3 activity against jasmonic acid amino acid conjugates and the elucidation of 12-O-β-glucopyranosyl-jasmonoyl-l-isoleucine as an additional metabolite. Phytochemistry 99:6–13. https://doi.org/10.1016/j.phytochem.2013.12.019
19. Bruckhoff V, Haroth S, Feussner K, König S, Brodhun F, Feussner I (2016) Functional characterization of CYP94-genes and identification of a novel jasmonate catabolite in flowers. PLoS One 11(7):e0159875. https://doi.org/10.1371/journal.pone.0159875
20. Widemann E, Heitz T, Miesch L, Miesch M, Heinrich C, Pinot F, Lugan R (2015) Identification of the 12-oxojasmonoyl-isoleucine, a new intermediate of jasmonate metabolism in Arabidopsis, by combining chemical derivatization and LC-MS/MS analysis. Metabolomics 11(4):991–997. https://doi.org/10.1007/s11306-014-0754-7
21. Iven T, König S, Singh S, Braus-Stromeyer SA, Bischoff M, Tietze LF, Braus GH, Lipka V, Feussner I, Dröge-Laser W (2012) Transcriptional activation and production of tryptophan-derived secondary metabolites in Arabidopsis roots contributes to the defense

against the fungal vascular pathogen *Verticillium longisporum*. Mol Plant 5(6):1389–1402. https://doi.org/10.1093/mp/sss044

22. Göbel C, Feussner I (2009) Methods for the analysis of oxylipins in plants. Phytochemistry 70:1485–1503. https://doi.org/10.1016/j.phytochem.2009.07.040
23. Wang L, Halitschke R, Kang JH, Berg A, Harnisch F, Baldwin IT (2007) Independently silencing two JAR family members impairs levels of trypsin proteinase inhibitors but not nicotine. Planta 226(1):159–167. https://doi.org/10.1007/s00425-007-0477-3
24. Glauser G, Dubugnon L, Mousavi SAR, Rudaz S, Wolfender J-L, Farmer EE (2009) Velocity estimates for signal propagation leading to systemic jasmonic acid accumulation in wounded Arabidopsis. J Biol Chem 284:34506–34513. https://doi.org/10.1074/jbc.M109.061432
25. Balcke GU, Handrick V, Bergau N, Fichtner M, Henning A, Stellmach H, Tissier A, Hause B, Frolov A (2012) An UPLC-MS/MS method for highly sensitive high-throughput analysis of phytohormones in plant tissues. Plant Methods 8(1):47. https://doi.org/10.1186/1746-4811-8-47

Chapter 14

Detection of Fungal Jasmonates by Liquid Chromatography Paired with Mass Spectrometry

Yingyao Liu, Rajesh N. Patkar, and Naweed I. Naqvi

Abstract

Liquid chromatography–mass spectrometry (LC-MS) is one of the most important analytical chemistry techniques for the detection and characterization of biologically active compounds of low abundance—for example, hormones. Gas chromatography (GC) coupled with mass spectrometry has been a method of choice to detect jasmonic acid, the well-known defense hormone in plants. Recently, we identified structural and functional analogs of phytohormone jasmonic acid (JA) and its derivatives, in the rice-blast fungus *Magnaporthe oryzae*. Here, we describe protocols of LC-MS/MS-based identification and quantification of fungal jasmonates, especially during pathogenic development in the rice blast fungus.

Key words Liquid chromatography, Small molecule, Mass spectrometry, Jasmonates, Oxylipin, Secondary metabolites, Fungal pathogen

1 Introduction

Liquid chromatography–mass spectrometry (LC-MS) is the most common and popular method of chemical analysis widely applied in both academic and industrial research studies [1, 2]. Usually, the first step during such chemical analyses involves chromatography, which separates individual compounds from a given mixture. The major difference between LC (liquid chromatography) and GC (gas chromatography) is the mobile phase—while an organic/ aqueous solvent is used in LC, a high-purity gas is used as a mobile phase in GC. Typically, GC-MS is most suited for volatile chemical compounds, however, due to the high temperature at which the method is performed, only a limited number of molecules can be analyzed by this technique [3]. Although the costs of operation and machine maintenance are high in LC, the major advantage that the LC offers is that the analysis is done at room temperature, a much more suitable condition for stability and function of a majority of chemical compounds. Furthermore, MS is a highly sensitive analytical method that detects molecules or ions based on their mass-to-

Antony Champion and Laurent Laplaze (eds.), *Jasmonate in Plant Biology: Methods and Protocols*, Methods in Molecular Biology, vol. 2085, https://doi.org/10.1007/978-1-0716-0142-6_14,

charge ratio (*m/z*; [4]). A typical mass spectrometer consists of three major components—an ion source where the molecule is ionized and fragmented, a mass analyzer or the dispersion system that separates the ions by their mass/charge ratio, and a detector, which measures the relative abundance of each ion [4].

Once individual compounds are separated in LC, they need to be ionized before being analyzed on the mass spectrometer, as only the ions, not neutral molecules, are detected by the instrument. Among various ionization methods, the commonly used techniques include electrospray ionization (ESI), atmospheric pressure chemical ionization (APCI), and electron impact (EI) [4, 5]. During ionization, two kinds of ions can be generated: a negatively charged ion created by giving an electron to the molecule or a positively charged product by taking an electron away from the molecule. Different analyzers are considered for different purposes—a quadrupole system, such as 'Quadrupole Time-of-Flight' (Q-ToF) for nontargeted metabolite profiling or global metabolome, a triple quadrupole for quantitative analysis, and ion traps (e.g., Orbitrap) for accurate mass detection or structural characterization of a molecular species from a given sample [6]. Data can be acquired from multiple product ions corresponding to one or more selected precursor ions using a 'Multiple Reaction Monitoring' (MRM) system. Next, a scientific analysis software (e.g., Agilent MassHunter work station) helps in instrument control, data acquisition, analyses and reporting of the data acquired.

In the field of metabolomics, a slight shift in the synthesis pathway of a molecule in vivo might result in a new derivative of the parental compound, leading to a significant biological change (functional gain or loss) inside the cell or organism. LC-MS is perhaps one of the most powerful techniques that can help detect/identify such molecules and/or their derivatives from a given cellular extract. One of the early reports on detection of jasmonic acid (JA) and related compounds in the culture filtrate of *Fusarium oxysporum* [7], as well as a recent article on detection of *cis*-Jasmone, for the first time, in *Lasiodiplodia theobromae* [8], described GC-MS-based identification of the jasmonates of fungal origin. Only methyl-JA and *cis*-Jasmone, among all the JA derivatives, are volatile and are suitable for detection by a GC-MS-based approach. With the help of LC-MS, we identified and quantified fungal jasmonate 12-hydroxy jasmonic acid (12OH-JA) for the first time in the rice-blast fungal pathogen *Magnaporthe oryzae* [9, 10]. Here, we describe the LC-MS/MS-based method used to detect and quantitate the levels of jasmonates (JA, MeJA and 12OH-JA) produced by *M. oryzae* during infection-related development.

2 Materials

The ultrapure water used in the preparation of solutions is deionized water with a sensitivity of 18 MΩ-cm at 25 °C. The organic solvents used are either analytical or MS-grade reagents purchased from Sigma Aldrich Corporation.

2.1 Cellular Extraction

1. Hydrophobic square glass sheets (80 × 80 mm or custom sized) (Deckglaser, Thermo Scientific).
2. Cell scraper (Corning Incorporated).
3. Sterile nylon membrane syringe filters (0.2 μm; Sartorius Minisart).
4. Cellular extraction buffer: deionized water/ethanol mixture (1:1).
5. Organic extraction buffer: methanol/chloroform mixture (1:1).

2.2 Chemical Analysis by LC-MS

1. AKTA Purifier (Amersham Biosciences).
2. Kinetex C18 (4.60 × 150 mm, 2.68 μ) reverse phase column (Phenomenex).
3. Zorbax Eclipse Plus C18 (2.1 × 50 mm, 1.8 μ) reverse phase column (Agilent Technologies).
4. Agilent Infinite 1290 UHPLC system (Agilent Technologies) for metabolite profiling on LC.
5. Accela 1250 LC system (Thermo Scientific) for separation of molecules for accurate mass detection on an orbitrap mass spectrometer.
6. Agilent ultra-high definition accurate-mass quadrupole time-of-flight (UHD Q-ToF) mass spectrometer for nontargeted metabolite profiling of crude samples or the partially purified fraction of interest.
7. LTQ Orbitrap Velos Pro Mass Spectrometer (Thermo Scientific) for accurate mass detection.
8. Agilent Triple Quad 6490 mass spectrometer (Agilent Technologies) for quantification.
9. LTQ Orbitrap Velos Pro mass spectrometer (Thermo Scientific).
10. MassHunter vB.05.01 work station (Agilent Technologies) and Xcalibur v2.2 software package (Thermo Scientific).
11. Mobile phase: Solvent A, deionized water with 0.1% formic acid; Solvent B, LC/MS-grade acetonitrile (ACN) with 0.1% formic acid solution.

3 Methods

Perform all the procedures at room temperature (25 °C) unless otherwise specified.

3.1 Sample Preparation from Fungal Cultures (Pathogenic Growth Phase)

1. Harvest conidia from 9- to 10-day-old *M. oryzae* culture and resuspend in sterile deionized water to get ~1×10^5 conidia per ml (*see* **Note 1**).
2. Inoculate droplets (~20–40 μl each) of conidial suspension on hydrophobic square glass sheets for conidial germination and appressorial development. Incubate the inoculated glass sheets under high humidity (one sheet each in a 150 mm sterile plastic petri dish) at 25 °C for 24 h (*see* **Note 2**).
3. At the end of incubation, collect the aqueous liquid on the glass sheet surface and store as total extracellular fluid (EF) at −80 °C until further processing. Lyse the appressoria adhering to the glass sheets by scraping the surface with a cell scraper. Hold the glass sheets in an inclined position, add 1 ml of 50% ethanol to the top end of the sheet and collect the resultant cell extract from the lysed appressoria in a 50 ml Falcon tube kept on ice. Store this intracellular extract as total appressorial extract (AE) at −80 °C until further processing.

   Lyophilize both the EF and AE completely and then add ~1 ml of organic extraction buffer for every 30 ml of EF or AE used in lyophilization to dissolve and extract the total metabolites. Keep the tubes on a tube roller at 25 °C for 2–3 h for better solubility and extraction of the molecules in the solvent mixture used. Filter the resultant organic extract by passing it through a 0.2 μm sterile nylon membrane syringe filter to get rid of cell debris and any insoluble compounds. Store at −80 °C until further use (*see* **Note 3**).

3.2 Chemical Analysis

3.2.1 Detection of Fungal Jasmonates

1. To isolate a jasmonate of interest, using a cell-based assay, run the total extracts (AE or EF) on a Kinetex C18 (4.60 × 150 mm, 2.68 μ) reverse phase column attached to an FPLC (AKTA Purifier) or HPLC (Agilent 1200 series) equipped with a fraction collector (Fig. 1). Set up a method for isocratic elution using 30% acetonitrile and a flow rate of ≤0.5 ml/min (*see* **Note 4**). Collect the individual fractions and test them using an appropriate cell-based assay—e.g., rice sheath inoculated with *M. oryzae* conidia or any relevant assay optimized earlier using the crude AE/EF (*see* **Note 5**) to narrow down to one or two purified fractions containing the jasmonate of interest. If required, the fraction(s) identified here can be further fractionated using another method such as size-exclusion chromatography and an aqueous mobile phase.

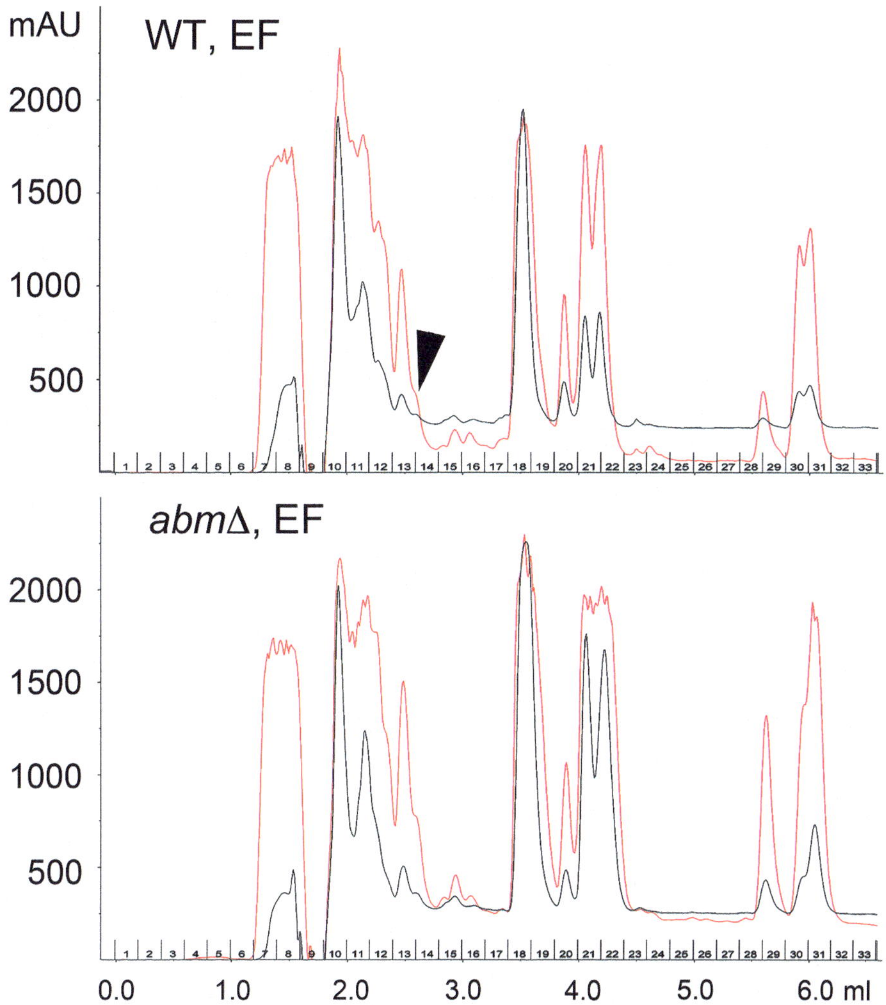

Fig. 1 LC profiles of total extracellular fluids (EF) of both the wild-type (WT; upper panel) and a knock-out strain (*abm*Δ; lower panel) of the rice-blast fungal pathogen *Magnaporthe oryzae*. The arrowhead marks the fraction (#13) of the WT EF containing the desired activity in a cell-based assay (rescue of *abm*Δ mutant). Chromatograms in red and black depict absorbance at 196 and 220 nm, respectively. (Reproduced from ref. 9 with permission from Springer Nature Limited)

Continue with the fraction containing the highest biological/functional activity for further molecular analysis.

2. Run the identified fraction from the aforementioned chromatographic separation(s) for a nontargeted metabolite profiling using Zorbax Eclipse Plus C18 column attached to Agilent Infinite 1290 UHPLC or an equivalent system mated to an UHD accurate-mass Q-ToF or an equivalent mass spectrometer.

3. Set the temperature of the column and auto sampler to 50 and 5 °C, respectively. Inject 5 μl of the sample and run with a flow rate of 0.3 ml/min. Set the gradient elution as follows:
   (a) Isocratic elution with 5% solvent B for 1.5 min;
   (b) Gradient to 98% solvent B in 10 to 12 min, hold at 98% solvent B for 15 min;
   (c) Return to 5% solvent B, wash the column and re-equilibrate in 16 min.
4. Ionize the molecules separated by the LC, using either positive or negative mode of ESI, with the following instrumental parameters:
   (a) Capillary voltage: 3600 V.
   (b) Collision energies: 10 V, 20 V and 40 V.
   (c) Fragmentor voltage: 115 V.
   (d) Nebulizer gas flow: 40 psi.
   (e) Drying gas flow: 10 l/min.
   (f) Sheath gas flow: 12 l/min.
   (g) Sheath gas temperature: 350 °C.
   (h) Source gas temperature: 250 °C.

Use Agilent MassHunter work station or equivalent to acquire and analyze the data and present it as extracted ion chromatogram (EIC) and base peak chromatogram (BPC) (Fig. 2, **Notes 6–8**).

Use LTQ Orbitrap Velos Pro mass spectrometer (Thermo Scientific) integrated with an Accela 1250 LC system to acquire accurate mass-to-charge ratio of the ion, and thus to identify the molecule with significant accuracy. Set the running conditions as follows:

(a) Resolution 60,000.
(b) AGC 5×10^5.
(c) Source voltage 3.5 kV.
(d) Capillary and source heater temperatures maintained at 300 °C with sheath gas, auxiliary gas, and sweep gas flow rates of 40 AU, 5 AU, and 1 AU, respectively.

Use collision induced dissociation (CID) with normalized collision energy of 35% and isolation width of 1 Da and AGC of 5×10^4 to get the MS/MS fragments of a parental ion of interest. Acquire and analyze the data using the Xcalibur v2.2 software package.

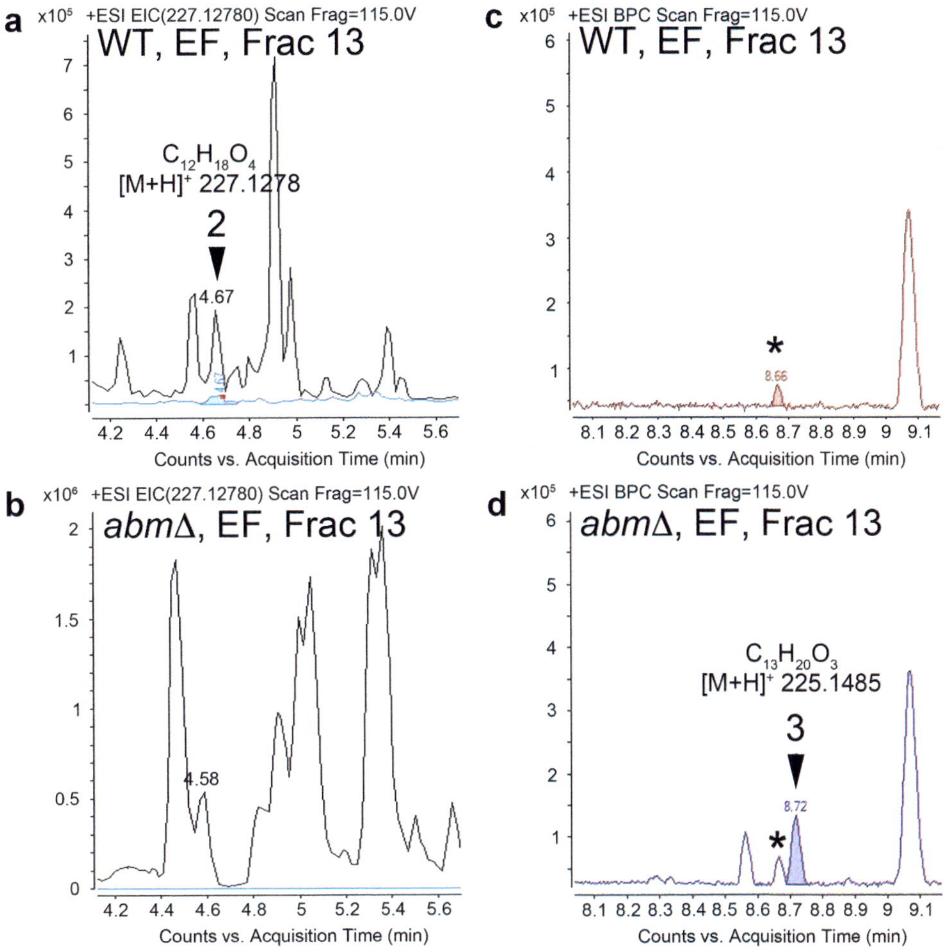

Fig. 2 Analysis of the fraction of interest from the WT EF (**a**, **c**) or the corresponding fraction from the *abm*Δ EFs (**b**, **d**) by mass spectrometry. Molecular ions differentially accumulated in the WT or *abm*Δ EF are depicted with arrowheads. The chromatographic peak area of the respective molecules in the peaks are shaded in blue. The background peak at retention time 8.66 min is denoted with an asterisk. (Reproduced from ref. 9 with permission from Springer Nature Limited)

3.2.2 Quantification of Jasmonate Levels

1. Prepare serial dilutions (1 nM to 10 μM) of the standard jasmonate of interest. Use a multiple reaction monitoring system (MRM) on an Agilent Triple Quad 6490 or an equivalent mass spectrometer to estimate the levels of the jasmonate-of-interest in the fraction identified, or in a given biological sample. First, separate the compounds using Zorbax Eclipse Plus C18 (2.1 × 50 mm, 1.8 μ) or equivalent reverse phase column at 50 °C and the aforementioned solvent A and solvent B as mobile phase. Inject 10 μl of the serially diluted standard or sample and run with a flow rate of 0.4 ml/min. Set up parameters for gradient elution as follows:

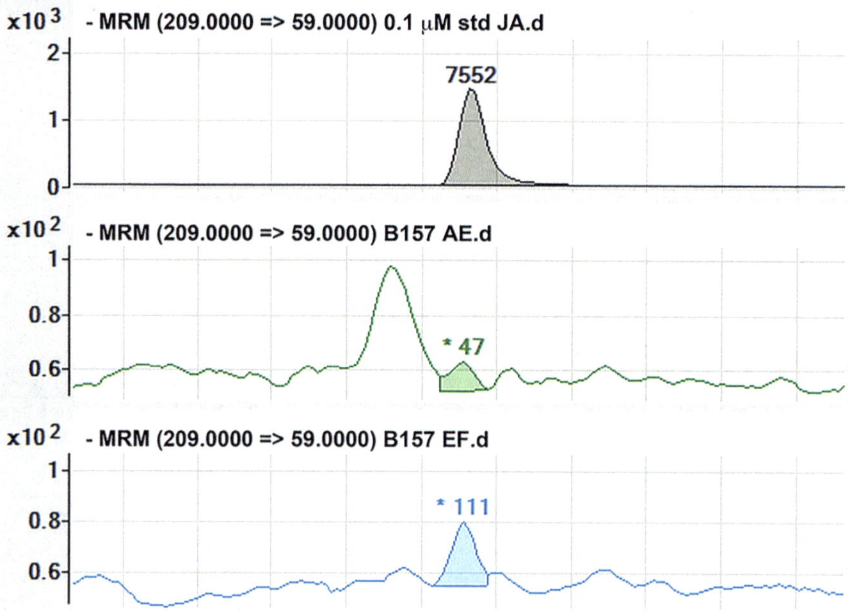

Fig. 3 Levels of jasmonic acid (JA) produced by *M. oryzae* at 24 h post inoculation (time of host invasion) estimated with the help of MRM signals from LC-MS/MS of JA accumulated extracellularly (EF; Extracellular Fluid) or intracellularly (AE; Appressorial Extract) in the wild-type fungal pathogen. (Reproduced from ref. 9 with permission from Springer Nature Limited)

(a) Isocratic elution with 5% solvent B for 1 min;

(b) Gradient to 98% solvent B in 6 min, hold for 10 min;

(c) Return to 5% solvent B, wash the column and re-equilibrate in 10 min.

2. Ionize the compounds using the ESI mode and the following settings on the instrument:

(a) Capillary voltage: 4000 V (+)/3000 V (−);

(b) Nebulizer gas flow: 40 psi.

(c) Drying gas flow: 14 l/min.

(d) Sheath gas flow: 12 l/min.

(e) Sheath gas temperature: 350 °C.

(f) Source gas temperature: 250 °C.

Set MRM transitions (precursor/parental and product ions) and collision voltage for different jasmonates as follows: JA (209 → 59, −25 eV); 12OH-JA (225 → 59, −25 eV) and MeJA (225 → 151, +15 eV).

Obtain and analyze the data using MassHunter vB.05.01 and estimate the levels of fungal jasmonate with the help of the analytical standard (Fig. 3).

4 Notes

1. Conidia should be freshly harvested, or the conidial suspension should be stored at 4 °C and used within a day of the preparation.
2. Scale up the biological sample preparation to a very high extent, in order to get the jasmonate of interest in the detectable range during mass spectrometric analysis.
3. Reconstitution of the lyophilized fractions in a smaller-than-original volume of an appropriate solvent is optional, but highly recommended for obtaining a concentrated sample for analysis.
4. Isocratic elution in HPLC at times may not show a good separation/resolution of the molecules. Therefore, a linear- or step-gradient elution may be considered to improve the resolution.
5. Before a cell-based assay is performed, lyophilize the fractions completely to remove the 30% acetonitrile used as the mobile phase. Resuspend the dried fraction content in deionized water and use an appropriate volume of the aqueously reconstituted fraction in the desired cell-based assay. The solvent (30% acetonitrile) alone should also be lyophilized and resuspended in water and should be used as a residual solvent control.
6. Extracted ion chromatogram is one of the chromatograms that one or more *m/z* values (interested analytes) are extracted from the entire data set of a chromatogram run. Base peak chromatogram, however, focuses on the most intense peak in each spectrum, which means that the background noises can be reduced dramatically. When doing data analysis, different kinds of chromatograms should be considered in a case-by-case manner [3, 11].
7. Although a standard (purified) compound can be detected in both positive and negative ionization modes, the same molecule from a biological sample may show up only in one of the two modes. For example, chemically synthesized 12OH-JA was detected in both positive and negative ionization modes, however, that from the *M. oryzae* sample was evident only in the negative mode.
8. Leaching from the surface of the glass vials used in mass spectrometry leads to addition of sodium ion to the molecule being ionized and consequently, to the detection of a sodium adduct (*m/z* of the parental ion +23 *m/z* of Na^+) rather than the parental compound per se. Therefore, it is highly recommended to scan for the requisite sodium adduct too during the analysis.

Acknowledgments

We thank the Temasek Life Sciences Laboratory (TLL Singapore) and the National Research Foundation (PMO Singapore, NRF-CRP7-2010-02 and NRF-CRP16-2015-04) for financial support. RP is supported by the DBT-Ramalingaswami Fellowship (BT/RLF/Re-entry/32/2014; Department of Biotechnology, Government of India).

References

1. Zhou B, Xiao JF, Tuli L, Ressom HW (2012) LC-MS-based metabolomics. Mol BioSyst 8:470–481
2. De Hoffmann E (2000) Mass spectrometry. Kirk-Othmer encyclopedia of chemical technology. John Wiley & Sons, Hoboken, NJ
3. Dailey AL (2017) LC-mass spectrometry for metabolomics. In: Espina V (ed) Molecular profiling. Methods in molecular biology, vol 1606. Humana Press, New York, NY
4. Nicolescu TO (2017) Interpretation of mass spectra. In: Aliofkhazraei M (ed) Mass spectrometry. IntechOpen Limited, London
5. Wilm M (2011) Principles of electrospray ionization. Mol Cell Proteomics 10(7):1–8
6. El-Aneed A, Cohen A, Banoub J (2009) Mass spectrometry, review of the basics: electrospray, MALDI, and commonly used mass analyzers. Appl Spectrosc Rev 44(3):210–230
7. Miersch O, Bohlmann H, Wasternack C (1999) Jasmonates and related compounds from *Fusarium oxysporum*. Phytochemistry 50 (4):517–523
8. Matsui R, Amano N, Takahashi K, Taguchi Y, Saburi W, Mori H, Kondo N, Matsuda K, Matsuura H (2017) Elucidation of the biosynthetic pathway of cis-jasmone in *Lasiodiplodia theobromae*. Sci Rep 7(1):6688–6697
9. Patkar RN, Benke PI, Qu Z, Chen YYC, Yang F, Swarup S, Naqvi NI (2015) A fungal monooxygenase-derived jasmonate attenuates host innate immunity. Nat Chem Biol 11 (9):733–743
10. Patkar RN, Naqvi NI (2017) Fungal manipulation of hormone-regulated plant defense. PLoS Pathog 13(6):e1006334
11. Murray KK, Boyd RK, Eberlin MN, Langley GJ, Li L, Naito Y (2013) Definitions of terms relating to mass spectrometry. Pure Appl Chem 85(7):1515–1609

Chapter 15

CRISPR/Cas9-Mediated Gene Editing of the Jasmonate Biosynthesis *OsAOC* Gene in Rice

Trang Hieu Nguyen, Huong To Thi Mai, Daniel Moukouanga, Michel Lebrun, Stephane Bellafiore, and Antony Champion

Abstract

The function of Jasmonate (JA) is well documented in different plant physiological processes as well as in the interactions with their environment. Mutants impaired in JA production and/or signaling are useful materials to study the function of this phytohormone. Genes involved in the JA biosynthesis pathway in rice have been described, but few mutants deficient in JA production and signaling have been identified. Moreover, these mutants are mostly generated through random mutagenesis approaches, such as irradiation, EMS treatment, or T-DNA insertion, and potentially harbor undesired mutations that could affect other biological processes. The CRISPR/Cas9 system is a precise and efficient genome editing tool that creates DNA modification at specific loci and limit undesired mutations.

In this chapter, we describe a procedure to generate new JA-deficient mutant using CRISPR/Cas9 system in rice. The *Allene Oxide Cyclase (OsAOC)* gene is targeted since it is a single copy gene in the JA biosynthesis pathway in rice. The widely used variety Oryza sativa *japonica* Kitaake has been chosen due to its short life cycle and its ease of genetic transformation. This protocol describes the selection of the 20-nt target sequence, construction of the binary vector, and strategy for selecting the T-DNA-free mutant.

Key words Jasmonate, Mutant, Rice, CRISPR/Cas9

1 Introduction

Rice (Oryza sativa) is one of the most important crops in the world and also a precious model for molecular genetic studies in monocotyledonous plant species. Understanding the molecular mechanisms of rice plant response to environmental constraints is essential for food security, especially in the context of climate change. Jasmonate is a lipid-based phytohormone that plays crucial roles in plant growth, development, and adaptation to abiotic and biotic stresses. In rice, JA regulates flower development and fertility

Electronic supplementary material: The online version of this chapter (https://doi.org/10.1007/978-1-0716-0142-6_15) contains supplementary material, which is available to authorized users.

Antony Champion and Laurent Laplaze (eds.), *Jasmonate in Plant Biology: Methods and Protocols*, Methods in Molecular Biology, vol. 2085, https://doi.org/10.1007/978-1-0716-0142-6_15,

[1–3], senescence [4], and light-interaction [5]. JA also triggers plant immunity system in response to bacteria [6], fungi [1], nematode attack [7], and herbivores [8, 9]. JA is also involved in plant response to several environmental stresses, such as nutrient starvation [10, 11], salt [12], drought [13], and cold stress [14].

Enzymes involved in each catalytic step of JA biosynthesis pathway have been described [15]. In rice, the JA biosynthesis pathway is conserved; however, there is a difference in the number of homologues that participate in each reaction. The conversion of allene oxide to 12-oxophytodienoic acid (OPDA) step is encoded by a single copy gene *Allene Oxide Cyclase* (*OsAOC*). Enzymes in other steps are encoded by at least two genes. These enzymes might have redundant activities or play specific function in response to particular stimuli. However, few of them have been functionally characterized [16]. Up to now, seven mutants defective in five genes of five different steps in JA biosynthesis were identified. *cpm1* (coleoptile photomorphogenesis 1) was the first JA mutant identified in rice (cv Nihonmasari) [17]. *cpm1* harbors a point mutation in one of the two AOS homologues. Other mutants, *hebiba* and *cpm2,* were found to be defective in the *OsAOC* single copy gene in rice [1]. The *eg1* (extra glumes 1) is a mutant impaired in a plastid-targeted lipase that participates in JA biosynthesis [2]. Two other mutants, *ucgl* and *og1*, both defective in the *Reductase 7* (*OsOPR7*) gene, and the mutant of the *Jasmonate Resistant 1* gene (*osjar1*) are affected in the production of the bioactive form JA-Ile [3, 18, 19]. Besides, two mutants *oscoi1b* [20] and *eg2/osjaz1* [2], involved in JA perception were also identified. Although these mutants covered different steps in both synthesis and signaling pathways, they were identified by random mutagenesis approaches. *cpm1, cpm2*, and *hebiba* were generated by irradiation, while *eg1, eg2/osjaz1*, and *ucgl* were obtained from ethylmethanesulfonate (EMS) treatment, and *osjar1* and *oscoi1b* were identified from a retrotransposon and T-DNA insertion library, respectively. These mutagenic methods frequently induce random and multisited mutations, which potentially affect other biological processes and require time-consuming backcross to remove unwanted mutations.

Targeted sequence editing systems could be used to avoid these issues. These include zinc-finger nucleases (ZFNs) and transcription activator-like effector (TALE) nucleases (TALENs) [21]. Most recently, a new RNA-guided genome editing approach, using the bacteria type II Clustered Regularly Interspaced Short Palindromic Repeats (CRISPR)-CRISPR-associated protein 9 (Cas9) (CRISPR/Cas9) from Streptococcus pyogenes, has made genome editing easier than ever [22]. The Cas9 nuclease is directed by a short single guide RNA (gRNA or sgRNA) to generate double-strand breaks (DSB) at specific sites in chromosomal DNA, thereby introducing mutations at the DSB via non-homologous end-joining repair mechanism of the host cell [23]. With the Cas9 gene being codon optimized for plant, generating stably genome edited plant is achieved by simply specifying 20-nt targeting sequence within its guide RNA, constructing the Cas9 coding gene and the gRNAs in the same vector and transform into plant. CRISPR/Cas9 was first applied to edit genome of a crop plant in 2013 [24, 25], this technique is now widely used in functional studies and is a powerful tool to speedup crop breeding.

Here, we describe a protocol to edit a JA biosynthesis gene *OsAOC* in rice cv Kitaake to generate new JA-free line via CRISPR/Cas9 system. We describe how to design the 20-nt target sequence, the construction of the binary vector, and the strategy used to select the T-DNA-free mutant plants.

2 Materials

2.1 Plant Material

1. *Oryza sativa* L. ssp japonica Kitaake seeds.

2.2 Bacteria and Plasmid

1. Expression vector pOs-Cas9 [24] modified (*see* Fig. 1).
2. Donor vector pUC57-sgRNA containing the sgRNA scaffold (*see* Fig. 1) (*see* **Note 1**).
3. Chemically competent *E. coli*.
4. *Agrobacterium tumefaciens* strain EHA105.

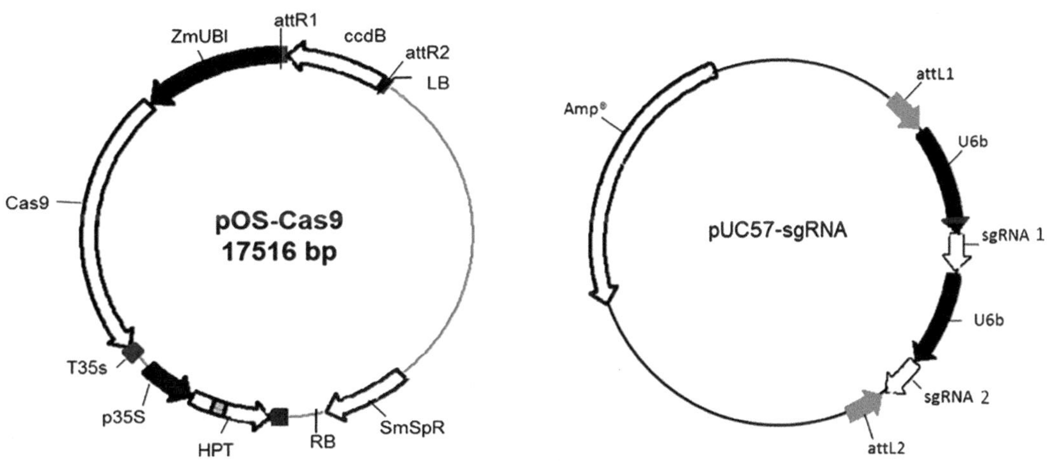

Fig. 1 Schematic diagram of the vectors used in this study. The codon-optimized Cas9 gene is cloned into the destination vector; an intron is inserted into the first exon of HPT gene to increase plant transformation efficiency. The whole cassette from *attL1* to *attL2* including two sgRNAs and its promoters is synthesized and cloned into entry vectors

2.3 Chemicals and Reagents

2.3.1 Binary Vector Construction and Bacterial Transformation

1. LR clonase enzyme.
2. Proteinase K.
3. Ethylenediaminetetraacetic acid (EDTA).
4. Plasmid miniprep kit.
5. Gel extraction kit.
6. Tris base, acetic acid, and EDTA (TAE) 10× buffer.
7. Spectinomycin–HCl pentahydrate.
8. LB broth medium: 10 g tryptone, 10 g sodium chloride, and 5 g yeast extract, 1 L H_2O, Adjust the pH to 7.0, autoclave.
9. LB agar plate: 10 g tryptone, 10 g sodium chloride, 5 g yeast extract, and 1 L H_2O. Adjust the pH to 7.0, add 15 g bacterial agar, and autoclave. Wait for media to cool down to about 60–65 °C, add 50 μg/mL spectinomycin and pour 20 mL per plate.

2.3.2 Selection of Transgenic Lines and Genotyping Mutation

1. DNA extraction buffer: 200 mM Tris–HCl pH 7.5, 250 mM NaCl, 25 mM EDTA, 0.5% SDS.
2. Liquid nitrogen.
3. DNA polymerase.
4. 70% and 100% ethanol.
5. Sterile ddH_2O.
6. 100 bp DNA ladder.
7. 1 kb DNA ladder.
8. Agarose.
9. TE 10× buffer.

2.3.3 Others

1. Microwave.
2. Incubator.
3. PCR thermocycler.
4. DNA electrophoresis apparatus.
5. DNA electroporation apparatus.
6. Benchtop microcentrifuge.
7. Plant transformation platform.
8. TissueLyser II.
9. NanoDrop.
10. CRISP-ID-an online tool for decoding CRISPR-mediated indels by Sanger sequencing (http://crispid.gbiomed.kuleuven.be/) [26].

3 Methods

3.1 Design gRNAs Targeting OsAOC Gene

We used the CRISPR Guides tool on Benchling platform (https://benchling.com/crispr) to design the gRNAs for CRISPR/Cas9 approach:

1. Download the full CDS sequence of *OsAOC* gene from The Rice Genome Annotation Project database (http://rice.plantbiology.msu.edu/analyses_search_locus.shtml) using the locus name LOC_Os03g32314.
2. Login the Benchling website; open a new CRISPR Guides project by clicking on the (+) button on the left toolbar→-CRISPR→CRISPR Guides.
3. Select the import form by raw bases, then copy and paste the CDS sequence.
4. Choose option Single guides, guides length = 20 base pairs, select target genome Oryza sativa *japonica* group→Finish.
5. Select the target region on the sequence map and press (+) button.
6. From the list of potential candidates, select two candidates with the best on-and-off-target score (*see* **Note 2**). Benchling gives 0–100 score for each candidate, higher score is better (an example of selection of target gRNAs that gave positive result in gene editing is shown in Table 1) (*see* **Note 3**).
7. After selection of gRNAs, the whole cassette containing the *attL1* and *attL2* sequences compatible with gateway cloning, the sgRNAs, and the pU6 promoter used to drive the transcription of sgRNAs are synthesized and cloned into the donor vector pUC57 (sequence is provided in Supplementary, plasmid map of pUC57-sgRNAs is shown in Fig. 1).

Table 1
Sequences of the two selected gRNAs

Position	Strand	Sequence	PAM	Specificity score	Efficiency score
20	−1	gCCTGACGGAGACTCGCGAGG	GGG	49.32196	66.903278
579	1	GGGCCCGTACCTGACCTACG	AGG	49.86528	68.600567

3.2 Plasmid Construction

1. Prepare LR Clonase Reaction on Ice:

Component	Volume
ddH_2O	1.5 μL
TE buffer 10×	0.5 μL
Destination vector pOs-Cas9	1 μL (160 ng)
Donor vector pUC57-sgRNAs	1 μL (160 ng)
LR Clonase II plus enzyme mix	1 μL
Total volume	5 μL

2. Mix well, spin down, and then incubate overnight at room temperature (22 °C –25 °C).
3. On the next day, stop LR Clonase reaction by adding 0.5 μL of Proteinase K into LR reaction and incubate at 37 °C for 10 min.
4. Use 2 μL product of LR Clonase reaction to transform *E. coli* chemically competent cell.
5. Growth 2–3 colonies overnight in 5 mL LB broth supplemented with 50 μg/mL Spectinomycin.
6. Perform Miniprep to isolate plasmid from overnight culture.
7. Verify pOs-Cas9-sgRNA plasmid constructs by Sanger sequencing using two primers pOs-Cas9-839 and pOs-Cas9-840 (*see* Table 2) located at two borders of *attR1* and *attR2* on the destination plasmid backbone.

3.3 Rice Transformation

1. Introduce the pOs-Cas9-gRNA plasmid into *Agrobacterium* strain EHA105 competent cells by electroporation method (*see* **Note 4**).
2. Transform rice embryogenic callus using *Agrobacterium*-mediated transformation method, many protocols for rice transformation are available (e.g., [27]). Try to keep as much as possible individual transformation events for further genotyping steps (*see* **Note 5**).

3.4 Genotyping Edited Transgenic Lines

The principle to have mutation without insertion is based on the independent segregation of T-DNA and the target loci following the Mendelian law. The strategy to select stable mutant without T-DNA insertion and purpose of each PCR are illustrated in Fig. 2 (*see* **Note 6**). Two PCR reactions have to be performed at T0 (PCR 1 and 2) and T1 (PCR 3 and 4) generations; primer sequences are shown in Table 2. Two primers, pUBI-F and Cas9-R, are designed

Table 2
List of primers

Name	Sequence (5′-3′)	Utility
pOs-Cas9–839 pOs-Cas9–840	AGGCGGGAAACGACAATCTG AGACATGCAATGCTCATTAT	Verify plasmid construction
Aoc-F Aoc-R	GGCATCACCAATCGGCAATC GTGAAGTTGTTGAGGCAGGC	Identify mutation
pUBI-F Cas9-R1216	GGATGATGGCATATGCAGCAG GAGTGTGAGGTCCTGGTGGT	Check the presence of T-DNA

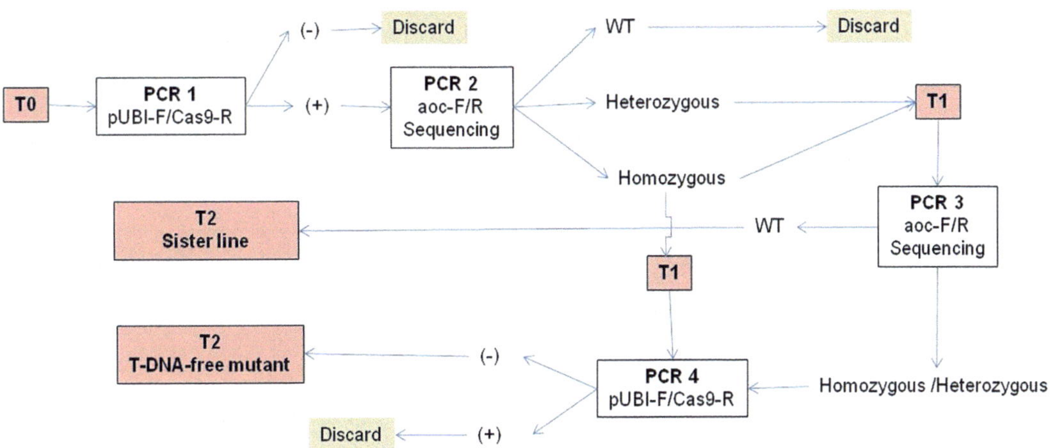

Fig. 2 Strategy to select T-DNA-free stable mutant from CRISPR/Cas9 method. PCR 1 and 4 use primers pUBI-F and Cas9-R. PCR 2 and 3 use aoc-F and aoc-R primers, and followed by Sanger sequencing PCR products

to detect the present of the *Cas9* gene. Product length is 1216 bp. To examine the mutations, a pair of gene-specific primers (aoc-F, 118 bp upstream from target 1 and aoc-R, 232 bp downstream from target 2, *see* Table 2) is designed to amplify the DNA segment flanking two targets sequence. Product length is 817 bp.

1. Sample preparation for genotyping:
   (a) Place about 20 mg of plant tissue in an Eppendorf tube.
   (b) Grind the sample with liquid nitrogen and TissueLyser II.
   (c) Add 400 μL of extraction buffer and mix by vortexing.
   (d) Centrifuge at >20,000 × *g* for 2 min.
   (e) Transfer the supernatant to a new Eppendorf tube (about 250 μL).
   (f) Add 1 volume of ice-cold isopropanol and incubate on ice for 30 min.

(g) Centrifuge in a microfuge at room temperature for 5 min.

(h) Discard the supernatant and dry the pellet.

(i) Resuspend in 100 μL of TE, estimate OD by NanoDrop and store at −20 °C.

2. Set up reactions for PCR amplification:

Component	Volume (50 μL reaction)	Final concentration
ddH_2O	22 μL	
5× GoTaq buffer	5 μL	1×
dNTP 10 mM	0.25 μL	0.1 mM
Forward primer 10 mM	0.5 μL	0.2 mM
Reverse primer 10 mM	0.5 μL	0.2 mM
Plant sample	1 μL	50–100 ng
GoTaq polymerase	0.1 μL	1 U

3. Amplify the targeted region by PCR:

Cycle step	Temperature	Time	Number of cycle
Initial denaturation	94 °C	5 min	1
Denaturation Annealing Extension	94 °C 56 °C 72 °C	30 s 30 s 60 s	35
Final extension	72 °C 4 °C	10 min Hold	1

4. Analyze the PCR products on agarose gel 2%.

5. For PCR 2 and 3, assess mutations by Sanger sequencing.

6. Analyze sequencing results by CRISPR-ID online tool and perform sequence alignment with reference sequence of *OsAOC* gene.

4 Notes

1. For convenience and precision, the full-sequence *attL2*-pU6b-sgRNA1-pU6b-sgRNA2-*attL1* of 1118 bp was synthesized and cloned into pUC57 as the donor vector. The whole gRNA scaffold is introduced into pH-UBI-Cas9 vector through *attR1* and *attR2* sequences compatible for gateway cloning method.

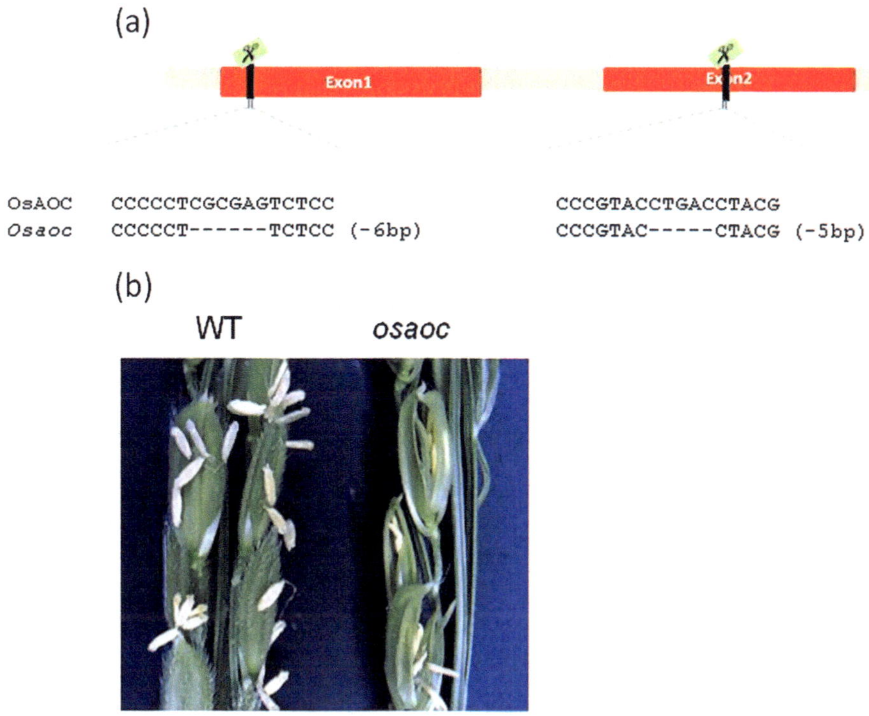

Fig. 3 (**a**) Genotype of *osaoc* mutant with two InDel at both exons; (**b**) Spikelet phenotype of *osaoc*

2. For guide sequences that do not begin with a G, it is recommended to add an additional G at the 5′ position of the guide sequence, which results in 21 bp of gRNA (which is the case of gRNA1 in this protocol, *see* Table 1). This additional G does not alter the specificity of the sgRNA or affect the efficiency of Cas9 cleavage.

3. The distance between two targets should about 100–500 bp for convenient genotyping of the mutation. One should balance between specific score and efficiency score, candidates with specificity and efficiency score above 50 should be considered.

4. It is important to verify the sequence of gRNA cassette before proceeding to rice transformation. Plasmid were isolated from *Agrobacteria* and sent for sequencing.

5. Note that *osaoc* mutant is male sterile (Fig. 3b). In order to multiply the mutant, it is recommended to maintain heterozygosity. The CRISPR/Cas9 system used in this procedure and the selection of double target gRNA lead to very high efficiency of gene modification (more than 90%). It is possible to select biallelic line that still produces seeds with one deficient allele. However, we recommend preparing a wild-type pollen donor plant to cross with the male sterile homozygous mutant. We

suggest sowing two seeds in a pot two times per week, continuously for 4 weeks and starting from regeneration step in plant transformation process.

6. We worked with about 80 individual lines at T0 generation. Later, we continued with 8–10 lines at T1 with either homozygous or heterozygous mutations and for each line we sowed at least 100 seeds.

References

1. Riemann M, Haga K, Shimizu T, Okada K, Ando S, Mochizuki S, Nishizawa Y, Yamanouchi U, Nick P, Yano M et al (2013) Identification of rice Allene oxide Cyclase mutants and the function of jasmonate for defence against Magnaporthe oryzae. Plant J 74:226–238
2. Cai Q, Yuan Z, Chen M, Yin C, Luo Z, Zhao X, Liang W, Hu J, Zhang D (2014) Jasmonic acid regulates spikelet development in rice. Nat Commun 5:3476
3. Liao L, Shi CH, Zeng DD, Jin XL, Wu JG (2015) Morphogenesis and molecular basis on the unclosed glumes, a novel mutation related to the floral organ of rice. Plant Mol Biol Rep 33:480–489
4. Fang C, Zhang H, Wan J, Wu Y, Li K, Jin C, Chen W, Wang S, Wang W, Zhang H et al (2016) Control of leaf senescence by an MeOH-jasmonates cascade that is epigenetically regulated by OsSRT1 in rice. Mol Plant 9:1366–1378
5. Svyatyna K, Riemann M (2012) Light-dependent regulation of the jasmonate pathway. Protoplasma 249(Suppl 2):S137–S145
6. Yamada S, Kano A, Tamaoki D, Miyamoto A, Shishido H, Miyoshi S, Taniguchi S, Akimitsu K, Gomi K (2012) Involvement of OsJAZ8 in jasmonate-induced resistance to bacterial blight in rice. Plant Cell Physiol 53:2060–2072
7. Nahar K, Kyndt T, De Vleesschauwer D, Höfte M, Gheysen G (2011) The jasmonate pathway is a key player in systemically induced defense against root knot nematodes in rice. Plant Physiol 157:305–316
8. Zhou G, Qi J, Ren N, Cheng J, Erb M, Mao B, Lou Y (2009) Silencing OsHI-LOX makes rice more susceptible to chewing herbivores, but enhances resistance to a phloem feeder. Plant J 60:638–648
9. Li R, Afsheen S, Xin Z, Han X, Lou Y (2013) OsNPR1 negatively regulates herbivore-induced JA and ethylene signaling and plant resistance to a chewing herbivore in rice. Physiol Plant 147:340–351
10. Kobayashi T, Itai RN, Senoura T, Oikawa T, Ishimaru Y, Ueda M, Nakanishi H, Nishizawa NK (2016) Jasmonate signaling is activated in the very early stages of iron deficiency responses in rice roots. Plant Mol Biol 91:533–547
11. Khan GA, Vogiatzaki E, Glauser G, Poirier Y (2016) Phosphate deficiency induces the jasmonate pathway and enhances resistance to insect herbivory. Plant Physiol 171:632–644
12. Ismail A, Seo M, Takebayashi Y, Kamiya Y, Eiche E, Nick P (2014) Salt adaptation requires efficient fine-tuning of jasmonate signalling. Protoplasma 251:881–898
13. Fu J, Wu H, Ma S, Xiang D, Liu R, Xiong L (2017) OsJAZ1 attenuates drought resistance by regulating JA and ABA signaling in rice. Front Plant Sci 8:2108
14. Du H, Liu H, Xiong L (2013) Endogenous auxin and jasmonic acid levels are differentially modulated by abiotic stresses in rice. Front Plant Sci 4:397
15. Wasternack C, Song S (2017) Jasmonates: biosynthesis, metabolism, and signaling by proteins activating and repressing transcription. J Exp Bot 68:1303–1321
16. Dhakarey R, Kodackattumannil Peethambaran P, Riemann M (2016) Functional analysis of jasmonates in rice through mutant approaches. Plants (Basel) 5
17. Biswas KK, Neumann R, Haga K, Yatoh O, Iino M (2003) Photomorphogenesis of rice seedlings: a mutant impaired in phytochrome-mediated inhibition of coleoptile growth. Plant Cell Physiol 44:242–254
18. Li X, Wang Y, Duan E, Qi Q, Zhou K, Lin Q, Wang D, Wang Y, Long W, Zhao Z et al (2018) Open Glume1: a key enzyme reducing the precursor of JA, participates in carbohydrate transport of lodicules during anthesis in rice. Plant Cell Rep 37:329–346

19. Xiao Y, Chen Y, Charnikhova T, Mulder PPJ, Heijmans J, Hoogenboom A, Agalou A, Michel C, Morel J-B, Dreni L et al (2014) OsJAR1 is required for JA-regulated floret opening and anther dehiscence in rice. Plant Mol Biol 86:19–33
20. Lee S-H, Sakuraba Y, Lee T, Kim K-W, An G, Lee HY, Paek N-C (2015) Mutation of *Oryza sativa* Coronatine insensitive 1b (OsCOI1b) delays leaf senescence. J Integr Plant Biol 57:562–576
21. Puchta H (2017) Applying CRISPR/Cas for genome engineering in plants: the best is yet to come. Curr Opin Plant Biol 36:1–8
22. Ran FA, Hsu PD, Wright J, Agarwala V, Scott DA, Zhang F (2013) Genome engineering using the CRISPR-Cas9 system. Nat Protoc 8:2281–2308
23. Jinek M, Chylinski K, Fonfara I, Hauer M, Doudna JA, Charpentier E (2012) A programmable dual-RNA-guided DNA endonuclease in adaptive bacterial immunity. Science 337:816–821
24. Miao J, Guo D, Zhang J, Huang Q, Qin G, Zhang X, Wan J, Gu H, Qu L-J (2013) Targeted mutagenesis in rice using CRISPR-Cas system. Cell Res 23:1233–1236
25. Shan Q, Wang Y, Li J, Zhang Y, Chen K, Liang Z, Zhang K, Liu J, Xi JJ, Qiu J-L et al (2013) Targeted genome modification of crop plants using a CRISPR-Cas system. Nat Biotechnol 31:686–688
26. Dehairs J, Talebi A, Cherifi Y, Swinnen JV (2016) CRISP-ID: decoding CRISPR mediated indels by sanger sequencing. Sci Rep 6
27. Hiei Y, Komari T (2008) *Agrobacterium*-mediated transformation of rice using immature embryos or calli induced from mature seed. Nat Protoc 3:824–834

Chapter 16

Mass Spectrometric Approaches to Study the Metabolism of Jasmonates: Biotransformation of Exogenously Supplemented Methyl Jasmonate by Cell Suspension Cultures of *Moringa oleifera*

Claude Y. Hamany Djande, Ntakadzeni E. Madala, and Ian A. Dubery

Abstract

Jasmonic acid (JA) and derivatives play a crucial role in plant adaptation to environmental stress. The exogenous application of methyl jasmonate (MeJA) results in activation of stress-related genes and subsequent production of secondary metabolites. This implies the biotransformation of the hormone into a more active form. In this study, *Moringa oleifera* cell suspension cultures were treated with 100 μM MeJA. Methanolic cell extracts were analyzed with reverse phase ultrahigh performance liquid chromatography (UHPLC) coupled to a quadrupole time-of-flight high definition mass spectrometer (QTOF–HDMS, with MS^E data acquisition). Using a targeted approach for jasmonate profiling, extracted ion chromatograms were generated. MS fragmentation patterns were subsequently derived and molecular formulae calculated from the MS data of each ion. Furthermore, triple quadrupole (QqQ) MS, operated in selected reaction monitoring (SRM) mode, was employed to target masses of interest. In addition to MeJA and JA, three jasmonoyl-amino acids were annotated: jasmonoyl-valine (JA-Val), jasmonoyl-isoleucine/leucine (JA-Ile/Leu), and jasmonoyl-phenylalanine (JA-Phe), based on characteristic precursor and product ions. Furthermore, JA conjugated to a hexose was observed, as well as hydroxylated and carboxylated derivatives of JA-amino acid conjugates. The data point to active metabolism of the externally added MeJA by the *M. oleifera* cells through biotransformation and bioconversion reactions that can be investigated in depth using advanced mass spectrometric analyses.

Key words Bioconversion, Jasmonic acid, Jasmonoyl-amino acids, Mass spectrometry, Methyl jasmonate, *Moringa oleifera*

1 Introduction

Plants have the ability to rapidly perceive and respond to biotic and abiotic environmental stresses by inducing secondary metabolites as an adaptive strategy [1]. Among the plant hormones, JA and derivatives (i.e., methyl jasmonate (MeJA and jasmonoyl-amino acids), collectively designated as jasmonates, are important regulators [2]. Jasmonates are fatty acid-derived compounds characterized

Antony Champion and Laurent Laplaze (eds.), *Jasmonate in Plant Biology: Methods and Protocols*, Methods in Molecular Biology, vol. 2085, https://doi.org/10.1007/978-1-0716-0142-6_16,

by a cyclopentanone structure [3, 4]. The molecules are synthesized through linolenic acid and hexadecatrienoic acid leading to the JA precursor, 12-oxo-phytodienoicacid (OPDA). Both abiotic and biotic stress stimuli can lead to JA production in plants. Upon stimulation, endogenous *in planta* levels of jasmonates can rise to the nanomolar range, with the timing and extent of the JA response dependent on the plant, tissue, stressor, and environment [4]. The biologically active form of JA is the isomer (+)-(3R,7R) (*cis*) that can be deactivated in vivo by epimerization to the inactive *trans* form (+)-(3R,7S) [5, 6]. However, the reversibility of the epimerization results in the presence of active JA at equilibrium.

Although JA was considered as the end product of the octadecanoic pathway and bioactive phytohormone, it has been established that some JA metabolites, derivatives, and biosynthetic precursors also exhibit biological activity [3]. JA-Ile ((+)-7-*iso*-JA-Ile) was reported to be the most bioactive compound in plant defense [3, 7]. Apart from JA-Ile, little is known regarding the bioactivity of other JA-amino acid conjugates and other JA derivatives in plants. It was suggested that they play a role in regulating the hormonal activity of JA.

Some modifications can occur to clear the accumulated JA in the cells, giving rise to the large group of jasmonates [5]. Modification reactions involve the methylation of JA to form MeJA, decarboxylation to form *cis*-jasmone, and *O*-glucosylation [4]. Hydroxylation, carboxylation, sulfatation, and conjugation with amino acids, such as Ile, Leu, Val, Gln, Trp, and Phe [4] have been reported.

In general, the hormonal activity of JA-amino acid conjugates is decreased by hydroxylation and oxidation [8]. JA-Ile levels can be controlled through hydrolysis of the amide bond resulting in JA and Ile [9, 10] or catabolized through ω-oxidation to generate 12OH-JA-Ile and 12COOH-JA-Ile [10, 11]. In *Arabidopsis thaliana,* for instance, deactivation of JA-Ile is accompanied by a shift toward oxidized derivatives to control hormone oxidation status for signal attenuation [10]. It was also proposed that JA-Ile is the metabolic hub that directs JA profile complexity and that the hormone oxidation ratio, rather than absolute changes in the levels of JA-Ile, is important in signaling.

Plant cells in suspension culture can serve as bioreactors and suitable material for the understanding of fundamental biosynthetic processes of secondary metabolites [12]. It has been demonstrated that exogenous application of relatively high concentrations of MeJA to different cell suspension cultures results in the production of terpenes [13], phenolics, and flavonoids [14]. *Moringa oleifera* is a tree with a rich phytochemical composition and reported biomedical activities [15]. Exogenous MeJA was applied to *M. oleifera* cell suspensions in order to evaluate the suitability of the inducer and the responsiveness of the cell system for secondary metabolite production.

Although jasmonates are key regulators in plants, the quantitative and qualitative profiling of jasmonates are still limited [5, 16–18]. Here, using ultrahigh performance liquid chromatography (UHPLC)–quadrupole time-of-flight high definition mass spectrometry (QTOF–HDMS) and triple quadrupole mass spectrometry (QqQ-MS), metabolites resulting from the bioconversion of MeJA in cell suspension cultures were detected and annotated.

2 Materials

2.1 Chemicals, Solutions, and Media

1. Standards: Methyl jasmonate (MeJA) and jasmonic acid (JA), Sigma, St Louis, MO, USA); (±)-jasmonic acid-isoleucine (JA-ileu) (Cayman Chemical).
2. Plant growth regulators/phytohormones: 2,4-dichlorophenoxy acetic acid (2,4-D; Sigma, St. Louis, MO, USA), kinetin (Kin, Duchefa, Haarlem, Netherlands) and alpha naphthalene acetic acid (NAA, Sigma, St Louis, MO, USA).
3. Murashige and Skoog (MS) medium [19]; Duchefa, Haarlem, Netherlands), supplemented with 3% sucrose and 0.1% hydrolyzed casein (Sigma, Munich, Germany).
4. Plant tissue culture facilities including laminar flow hood and orbital shaker.
5. Chromatography and mass spectrometry solvents: Methanol (MeOH) and acetonitrile, analytical/UHPLC grade (Romil SpS™, Cambridge, UK), formic acid, UHPLC grade (Merck, Darmstadt, Germany) (*see* **Note 1**).
6. Ultra-Turrax homogenizer (CAT Scientific, Paso Robles, CA, USA).
7. Rotary evaporator with heating bath.
8. Sterile plastic Falcon tubes (50 mL) and Eppendorf (microcentrifuge) tubes (2 mL).
9. 0.2 μm nylon syringe filters and chromatographic vials with inserts and screw caps.

2.2 Cell Suspension Cultures

1. *M. oleifera* liquid medium consisted of the MS medium to which the phytohormones 2,4-dichlorophenoxy acetic acid (113 μM), kinetin (116 μM), and alpha naphthalene acetic acid (107 μM) were added. The final pH was adjusted to 5.8.
2. Cell suspensions were initiated from 2 g friable callus (initiated from aseptically grown *M. oleifera* stem segments cultured on the medium described above) into Erlenmeyer flasks

containing 100 mL of the sterile liquid medium. The cells in suspension were incubated on an orbital shaker at 130 rpm and 25 °C and subcultured into fresh medium every 7 days for a period of 3 weeks to allow for adaptation to growth in liquid medium. Once successfully established, the suspensions were subcultured every 7 days.

2.3 Liquid Chromatography System

1. Ultrahigh performance liquid chromatography (UHPLC) system: For example, Acquity UHPLC system (Waters, Corporation, Milford, MA, USA) equipped with an Acquity UHPLC column (HSS T3 reverse phase—150 × 2.1 mm with a 1.8 μm particle size).

2.4 Mass Spectrometer Instrumentation and Software

1. High definition, accurate mass, quadrupole time-of-flight (qTOF) mass spectrometer (MS) with electrospray ionization (ESI): For example, a SYNAPT instrument (Waters Corporation, Milford, MA, USA), with MassLynx™ XS and MarkerLynx™ XS software (Waters Corporation, Milford, MA, USA).
2. UHPLC triple quadrupole (QqQ) MS, able to operate in a selected reaction monitoring (SRM) mode: For example, a Nexera 8030 instrument (Shimadzu, Kyoto, Japan) with Profiling Solution software (Shimadzu, Kyoto, Japan).

3 Methods

3.1 Cell Suspension Elicitation with MeJA

1. Transfer 20 mL of cell suspension, 3 days after subculture (corresponding to the early- to mid-log phase of the batch culture), into pre-weighed sterile 50 mL Falcon tubes. Work in a sterile laminar flow hood and add concentrated MeJA (4.3 M) to a final concentration of 100 μM (*see* **Note 2**) and incubate on an orbital shaker for 24 h. Non-treated samples serve as negative controls and are harvested at 0 h and 24 h to account for any variation not related to the treatment. Include three independent biological replicates with three technical replicates of each ($n = 9$) in the experimental design (*see* **Note 3**).

3.2 Metabolite Extraction and Sample Preparation

1. Harvest elicited cell suspensions by centrifugation in a benchtop swinging-bucket centrifuge at 5100 × *g* for 30 min at 4 °C and separate the media from the cells. Add 100% methanol to the pelleted cells (average mass 8.4 g) in a 1:1.5 (v/m) ratio and place on ice to quench enzymatic activities (*see* **Notes 4** and **5**). Homogenize cells for 1 min and centrifuge under the same conditions as above.
2. Carefully transfer the supernatants into round bottom flasks for concentration, using a rotary evaporator set at 55 °C.

Evaporate the MeOH extracts to about 1 mL and transfer to 2 mL Eppendorf tubes to dry to completeness in a heating block set at 55 °C under constant air flow. Reconstitute the dried residues in 50% (v/v) aqueous UHPLC-grade MeOH and filter through 0.22 μm nylon filters into chromatographic vials (*see* **Note 6**).

3.3 Ultrahigh Performance Liquid Chromatography

1. Analyze the methanolic extracts prepared from cultured cells in triplicate (*see* **Note 2**) using an UHPLC system combined to a high definition mass spectrometer (MS) detector. For separation of compounds/analytes, use a binary solvent mixture (mobile phase) consisting of water (eluent A) and acetonitrile (eluent B), both containing 0.1% of formic acid at a flow rate of 0.4 mL/min. Following the initial conditions of 2% B, kept constant for 1 min, inject 2 μL of the samples. Introduce a gradient to change the chromatographic conditions from 2 to 3% B over 1–3 min and then increase to 8% B at 4 min. Eluent B is further increased to 50% over 4–25 min and ramped to 95% in 1 min. The concentration of B is held constant at 95% from 26–27 min and restored to the initial condition of 2% B over 28–30 min.

3.4 Quadrupole Time-of-Flight High Definition Mass Spectrometry

1. Use the negative ESI mode of the qTOF–MS instrument in V-optics to further detect the UHPLC-separated analytes. Set the method to perform unfragmented and four fragmented experiments (MS^E) simultaneously by collision energy ramping from 10 to 40 eV (*see* **Note 7**. Leucine encephalin (50 pg/mL, 554.2615 Da) is used as calibrant with a mass accuracy window of 0.5 mDa. Operate the MS with capillary, sampling, and extraction cone voltages of 2.5 kV, 30 V, and 4 V, respectively, with Nitrogen as nebulization gas. Use the cone gas and dissolvation gas flows at 50 and 550 L/h, respectively, with the dissolvation and source gas temperatures at 450 and 120 °C, respectively. Set the interscan time as 0.02 s and the scan time at 0.2 s to cover a mass to charge (m/z) range of 100–1000.
2. Representative base peak intensity (BPI) (*see* **Note 8**) MS chromatograms (Fig. 1) Show differences in peak intensities as well as appearance and disappearance of some peaks (indicated by the orange rectangles). This clearly indicates that MeJA perturbed/altered metabolism within *M. oleifera* cells resulting in changes in the chromatographic profile. Moreover, the majority of differences are observed in the late retention (R_t) region of the chromatograms, indicating the non-polar nature of compounds resulting from the treatment.

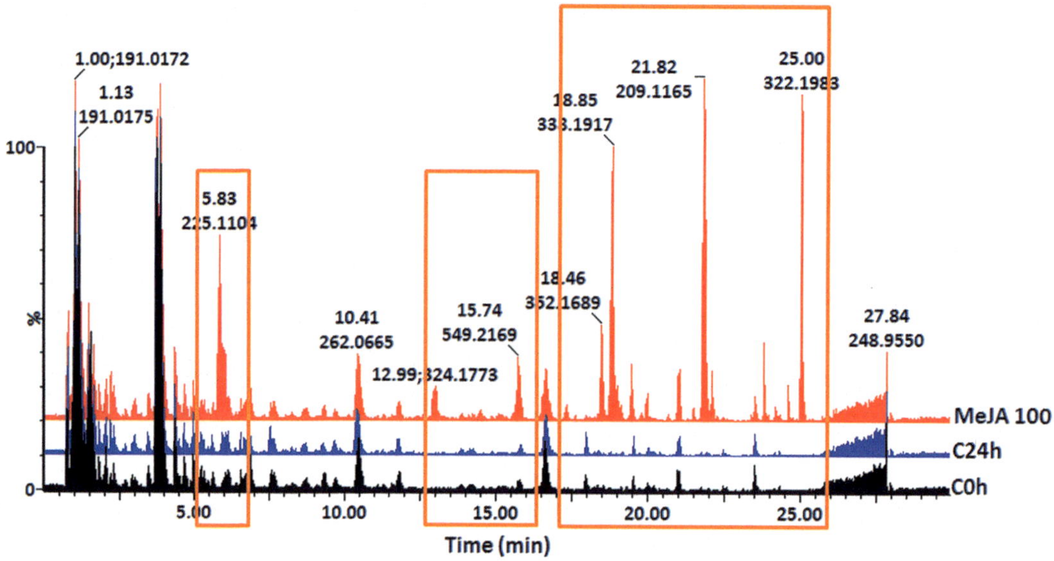

Fig. 1 UHPLC–MS (ESI negative mode) BPI chromatograms of methanol extracts from *Moringa oleifera* cells grown in suspension. Extracts were prepared in triplicate from harvested cells treated with 100 μM MeJA and incubated for 24 h. C0h is the non-treated control harvested before incubation and C24h represents the non-treated samples harvested after incubation

3.5 Triple Quadrupole (QqQ) Selected Reaction Monitoring (SRM) MS Analysis

1. As part of an orthogonal identification tool, a UHPLC QqQ MS, operated in an optimized SRM mode (*see* **Note 9**), is employed to target masses of interest. A reverse phase C_{18} analytical UHPLC column is used with a binary solvent mixture consisting of eluent A (0.1% formic acid in MilliQ water) and B (0.1% formic acid in acetonitrile) at a flow rate of 0.4 mL/min and injection volume of 4 μL. The elution is started with 5% of eluent B for 3 min. Gradient elution is then introduced to change the chromatographic conditions to 6% B (3–5 min) and then 11% B (5–6 min), respectively. The concentration of B is further increased to 50% over 6–27 min and kept constant for 2 min. The initial conditions are restored at 29 min and the column is allowed to re-equilibrate from 30 to 32 min.
2. For MS analysis, ESI in both positive and negative modes is used (the interface voltage at 4.50 kV, the interface current at 2.75 μA, the dissolvation and heat block temperatures at 249 °C and 400 °C, respectively). Argon is used as nebulizing gas at a flow rate of 1.50 L/min and Nitrogen as drying gas at a flow rate of 15.00 L/min. Target the JA-derived ions at *m/z* 209.10 in negative ionization mode and MeJA-derived ions at *m/z* 225.20 in positive ionization mode. Perform the analyses and acquire the data using the optimized fragmentation conditions for the transitions used in the SRM analysis: JA: [M-H], fragment ions: 209.10 > 59.00, 209.10 > 41.00; (Q1 = 10 V,

Collision energy = 13.00/39.00 V, Q3 = 10.00/13.00 V). MeJA [M+H]+, fragment ions: 225.20 > 151.15, 225.20 > 133.15, 225.20 > 147.15; (Q1 = −20.00 V, Collision energy = −15 V, Q3 = −20 V) (*see* **Note 9**).

3.6 Annotation of Metabolites

1. The annotation of JA-related compounds, extracted with 100% MeOH after treatment with MeJA, is performed using accurate mass data obtained from the QTOF–MS. Using a targeted approach, a putative identification (*see* **Note 10**) can be carried if no authentic standards (except for MeJA and JA) can be obtained. Obtain extracted ion chromatograms (XIC) for ions of interest and generate the MS fragmentation patterns. From the mass spectral data of each ion, molecular formulae can be calculated and submitted to online databases, such as Dictionary of Natural Products (DNP, http://dnp.chemnetbase.com), Knapsack (http://kanaya.naist.jp/KNApSAcK/) and ChemSpider (http://chemspider.com). The fragment ions can also be compared to previously published studies. As an additional structural analysis tool, potential structures matching the molecular formulae can be submitted to mass fragmentation analysis using MarkerLynx™ XS software (MassFragment tool) to confirm the fragmentation patterns of annotated compounds.

3.7 Characterization of Jasmonic Acid Derivatives

1. Jasmonic acid and methyl jasmonate:

   JA is a polyunsaturated fatty acid-derived molecule characterized by a pentacyclic ring structure (cyclopentanone) to which a carboxylic and aliphatic side chain is attached [20]. Two compounds (Table 1—(1, 2)), detected at R_t = 21.81 and 22.11 min with *m/z* of 209.114 (in negative mode), correspond to JA (Fig. 2a). However, with the QTOF–MS method, the fragmentation pattern of the extracted chromatogram reveals the presence of additional fragments of *m/z* 441 and 433 (Fig. 2b, c), which might be artifactual dimers of JA.

   In order to confirm the presence of JA in the extracts, analyze an authentic standard of JA with the optimized MRM method targeting *m/z* = 209.10 (JA in negative ionization mode) on the QqQ MS instrument to annotate the peaks as JA isomers (Fig. 3a, b). It is also possible that unidentified JA conjugates might co-elute with JA and hence interfere with the fragmentation pattern of the latter. This illustrates the power of tandem MS approaches (in this case MRM) to efficiently and selectively distinguish between isomers. Similarly, MeJA can be targeted (*m/z* = 225.20 in positive ionization mode) to detect three isomers (Fig. 3c, d) (*see* **Note 11**).

Table 1
Metabolites originating from jasmonic acid metabolism, identified from methanolic extracts of *Moringa oleifera* cell suspensions treated with 100 μM methyl jasmonate and incubated for 24 h

	Annotation	Molecular formulae	R_t (min)	MW	*m/z*	Diagnostic fragments	Instrument and ionization mode	Relative peak intensity
Methyl jasmonate and jasmonic acid								
1	Jasmonic acid (JA)	$C_{12}H_{18}O_3$	21.81 14.22	210	209.114	209	QTOF (−) QqQ	43.6 –
2	Jasmonic acid (JA)	$C_{12}H_{18}O_3$	22.11 14.50	210	209.113	209	QTOF (−) QqQ (−)	15.1 –
3	Methyl jasmonate (MeJA)	$C_{13}H_{20}O_3$	19.31, 19.60, 19.92	225	224.296	151, 147, 133	QqQ (+)	–
Jasmonoyl-amino acid conjugates								
4	Jasmonoyl-isoleucine/leucine (JA-Ile/Leu)	$C_{18}H_{29}NO_4$	25.00	323	322.199	130	QTOF (−)	437.6
5	Dihydrojasmonoyl-isoleucine/leucine (DihydroJA-Ile)	$C_{18}H_{31}NO_4$	24.62	325	324.214	130	QTOF (−)	38.8
6	12-Carboxyjasmonoyl-isoleucine/Leucine (12COOH-JA-Ile/Leu)	$C_{18}H_{27}NO_6$	18.47	353	352.172	130	QTOF (−)	105.2
7	*Trans*-12-hydroxyjasmonoyl-isoleucine/leucine (12OH-JA-Ile/Leu)	$C_{18}H_{29}NO_5$	18.83	339	338.194	130	QTOF (−)	310.4
8	*Cis*-12-hydroxyjasmonoyl-isoleucine/leucine isomer II (12OH-JA-Ile/Leu)	$C_{18}H_{29}NO_5$	19.04	339	338.194	130	QTOF (−)	37.9
9	Jasmonoyl-valine (JA-Val)	$C_{17}H_{27}NO_4$	23.78	309	308.181	116	QTOF (−)	89.2
10	Hydroxyjasmonoyl-valine (OH-JA-Val)	$C_{17}H_{27}NO_5$	12.95	325	324.177	116	QTOF (−)	41.9

11	Jasmonoyl-phenylalanine (JA-Phe)	$C_{21}H_{27}NO_4$	25.16	357	356.183	164	QTOF (−)	6.6
12	Hydroxyjasmonoyl-phenylalanine (OH-JA-Phe)	$C_{21}H_{27}NO_5$	19.98	373	372.178	164	QTOF (−)	30.1
Jasmonic acid hexose and other jasmonic acid conjugates								
13	Jasmonic acid (JA) conjugate	–	15.74	540	539.216	503, 209	QTOF (−)	25.6
14	Jasmonic acid hexose (JA-hexose) conjugate isomer 1	–	16.55	418	417.172	407, 371, 209	QTOF (−)	26.8
15	Jasmonic acid hexose (JA-hexose) conjugate isomer 2	–	16.73	418	417.173	407, 371, 209	QTOF (−)	17.7
16	Dihydrojasmonic acid (H_2JA) conjugate 1	–	17.31	420	419.189	211	QTOF (−)	18.9
17	Dihydrojasmonic acid (H_2JA) conjugate 2	–	21.49	566	565.249	519, 355, 211	QTOF (−)	17.1

QTOF quadrupole time-of-flight mass spectrometer, *QqQ* triple quadrupole mass spectrometer, (−) negative ionization mode, (+) positive ionization mode
m/z values of the precursor ions and product ions are reported to 3 an 0 decimals, respectively

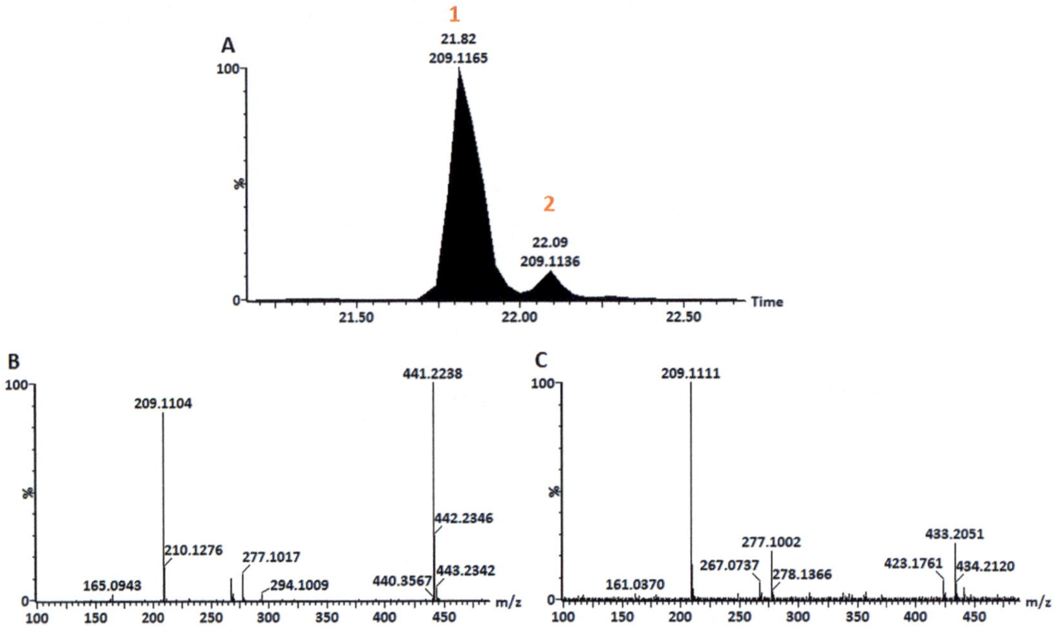

Fig. 2 (**a**) UHPLC–MS extracted ion chromatogram (XIC) of *m/z* 209 corresponding to JA; (**b**) mass spectrum corresponding to the base peak 1; and (**c**): mass spectrum of the base peak 2

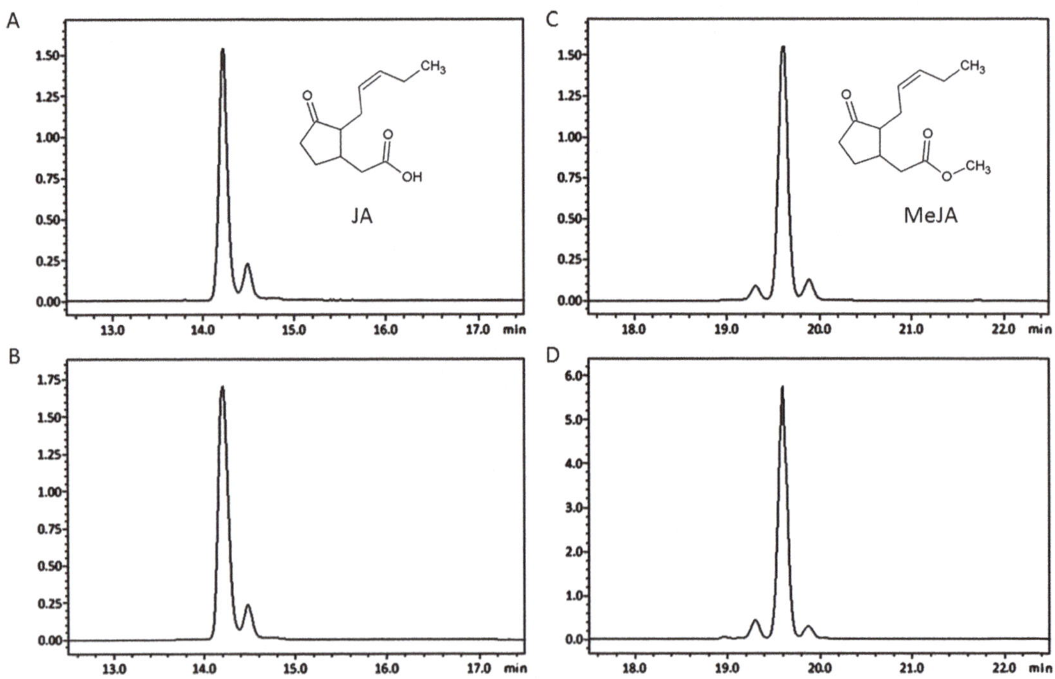

Fig. 3 UHPLC-single ion monitoring-MS chromatograms of JA and MeJA, obtained on a QqQ-MS. (**a**) JA standard; (**b**) JA in *Moringa oleifera* cell suspension sample treated with 100 μM MeJA; (**c**) MeJA standard; and (**d**) MeJA in extracts of cell suspension samples treated with 100 μM MeJA

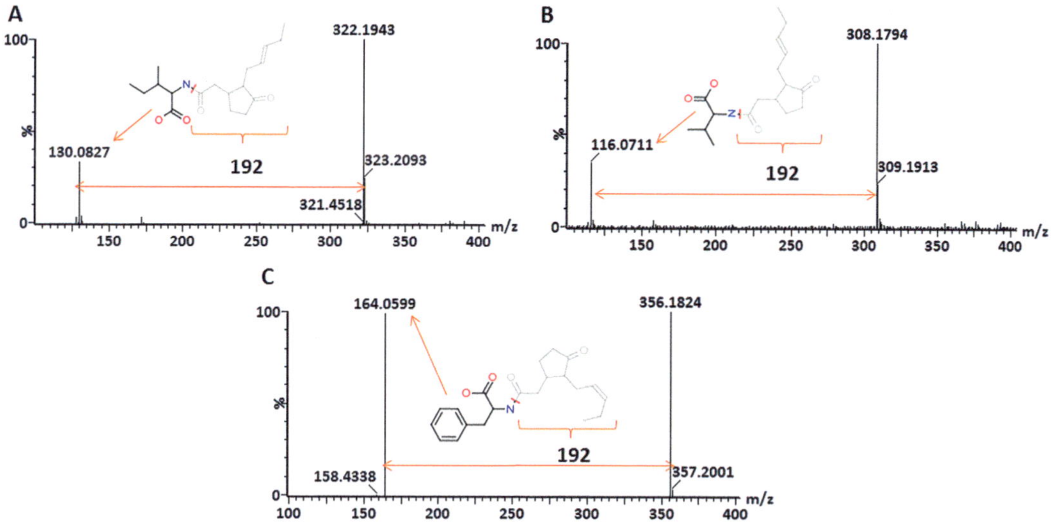

Fig. 4 Mass fragmentation patterns of JA-conjugated amino acids characterized in methanolic extracts prepared from MeJA-elicited *Moringa oleifera* cells. (**a**) JA-Ile; (**b**) JA-Val; and (**c**) JA-Phe. The annotations were supported using the MassFragment™ software as indicated in the figures. The arrows indicate the neutral loss of 192 corresponding to the JA without a hydroxyl group

2. Jasmonoyl-isoleucine (JA-Ile) and derivatives: A precursor ion *m/z* 322.199 elutes at $R_t = 25.00$ min, exhibiting a fragment ion of *m/z* 130, thus corresponding to isoleucine/leucine (Ile/Leu) (Table 1—(4)). The neutral loss of 192 corresponds to JA without a hydroxyl group ($[M–H–OH]^- = 210\text{-}1\text{-}17$). Based on the fragmentation pattern (Fig. 4a) and literature [21, 22], the compound can be annotated as jasmonoyl-isoleucine/leucine (JA-Ile/Leu). Although JA-Ile and JA-Leu can elute at the same R_t, the compound is more likely to be JA-Ile as it is the most prominent, bioactive, and studied among all the JAs [7]. Relatedly, a compound with a precursor ion of *m/z* 324.214 and a fragment ion of *m/z* 130 elutes at $R_t = 24.62$ min and can be assigned as dihydrojasmonoyl-isoleucine (H_2JA-Ile) (Table 1—(5)). The neutral loss between *m/z* 324 and 130 is 194, corresponding to dihydrojasmonic acid (H_2JA) without the hydroxyl group [23].

   Three compounds that share the same fragment ion of *m/z* 130 can be detected at R_ts = 18.47, 18.83, and 19.04 min with *m/z* 352.172, 338.194, and 338.194, respectively (Table 1—(6–8)). The last two compounds are isomers as these display the same mass and fragmentation patterns. The difference between *m/z* 352 and *m/z* 322 (JA-Ile) is 30, corresponding to two oxygens and two hydrogens (*m/z* 352 = *m/z* JA-Ile−2H + 2O). The addition of oxygens can be explained by the presence of a carboxylic acid moiety on the structure of *m/z* 352 and the loss of two hydrogens due to the formation of a double bond. In the case of *m/z* 338, the difference is one oxygen atom (*m/z* 338 = *m/z* JA-Ile + O). The compounds can thus be annotated as 12-carboxyjasmonoyl-isoleucine (12COOH-JA-Ile) and 12-hydroxyljasmonoyl-isoleucine (12OH-JA-Ile) isomers I and II, respectively [5, 8, 21]. Moreover, the two isomers of 12OH-JA-Ile correspond to the *trans* and *cis* isomers, respectively (Table 1—(7, 8)), as previously described [5].

3. Jasmonoyl-valine (JA-Val), jasmonoyl-phenylalanine (JA-Phe) and derivatives: A parent ion with *m/z* 308.181 is detected at R_t = 23.78 (Table 1—(9)). The derived fragment ion is *m/z* 116, corresponding to valine. The neutral loss of 192 is similar to the one observed with JA-Ile. The compound can thus be characterized as jasmonoyl-valine (JA-Val) (Fig. 4b). The hydroxylated derivative with *m/z* 324.177, giving a diagnostic product ion of *m/z* 116 and the mass difference between the JA-Val and OH-JA-Val of 16 (corresponding to an oxygen atom), can be annotated as hydroxyjasmonoyl-valine (OH-JA-Val) (Table 1—(10)) [21]. Similarly, jasmonoyl-phenylalanine (JA-Phe) (Fig. 4c) and hydroxyjasmonoyl-phenylalanine (OH-JA-Phe) correspond to the ions at *m/z* 356.183 (R_t = 25.16 min) and 372.178 (R_t = 19.98 min), respectively (Table 1—(11, 12)). In both cases, the fragment ion is *m/z* 164 matching that of phenylalanine [8, 21].

4. Jasmonic/dihydrojasmonic acid and derivatives: A compound of *m/z* 539.216 is detectable at R_t = 15.74 min (Table 1—(13)), with a daughter ion of *m/z* 209, corresponding to the *m/z* of JA. The compound is thus characterized as a JA conjugate since the *m/z* of the molecule conjugated to the JA was unknown.

   Two isomers with *m/z* 417.172 and 417.173 elute at R_t = 16.55 min and 16.73 min, respectively (Table 1—(14, 15)). The diagnostic fragment ions exhibit *m/z* 209 (JA), *m/z* 371 corresponding to JA-hexose (209 + neutral loss of 162 corresponding to a hexose moiety = 371) and *m/z* 407 (unknown). The compounds can accordingly be annotated as JA-hexose conjugate isomers I and II. Dihydrojasmonic acid (H_2JA) is characterized by *m/z* 211, different from JA (*m/z* 209) by two hydrogen atoms corresponding to the formation of a double bond [21, 22]. Two H_2JA conjugates with *m/z* 419.189 and 565.249 are detectable at R_t = 17.31 and 21.49 min, respectively (Table 1—(16, 17)). The diagnostic ions include *m/z* 211 corresponding to H_2JA. The nature of the molecules conjugated to JA, JA-hexose, and H_2JA are still to be investigated.

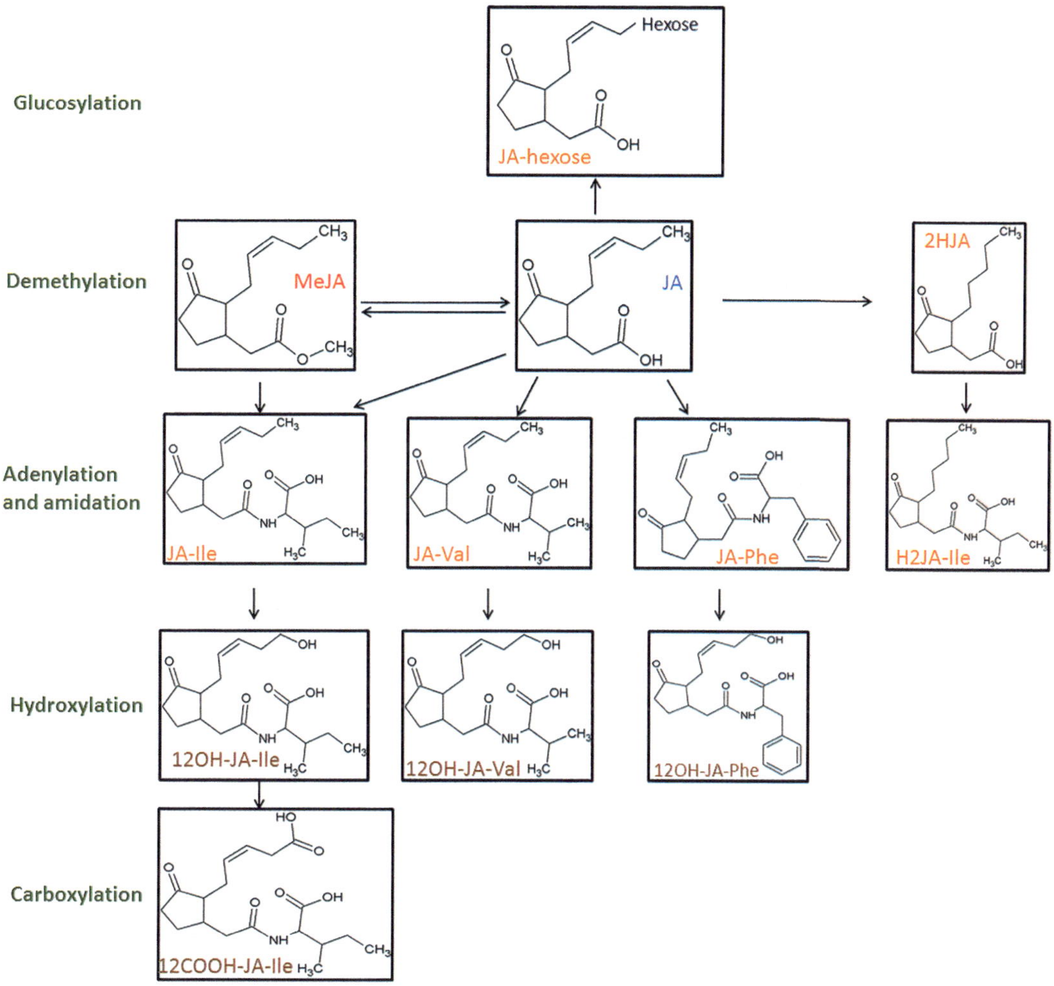

Fig. 5 Structures of annotated metabolites in Table 1 and metabolism of exogenously added MeJA to *Moringa oleifera* cell suspension cultures. Briefly, MeJA can be demethylated to form JA. Activation of the free carboxylic group of JA through adenylation allows for the condensation of the JA to amino acids (Ileu, Leu, Val, Phe) through the formation of an amide. JA can also be attached to glucose to form JA-glucosyl conjugates. Finally, JA-amino acids can be hydroxylated to form 12OH-JA-Ile, 12OH-JA-Val, and 12OH-JA-Phe and carboxylated to 12COOH-JA-Ile. JA derivatives can be considered to be storage molecules that may allow the rapid release of JA upon hydrolysis or/and involved in regulating the hormonal activity of JA

3.8 Observations

1. The application of MeJA to *M. oleifera* cells in suspension culture results in perturbation of the homeostasis within the cells, and triggers compensatory metabolic events that result in de-esterification MeJA to free JA (Fig. 5).
2. The judicious application of mass spectrometric analyses allows for the monitoring of biotransformation and bioconversion reactions that lead to the synthesis and subsequent metabolism of the identified Jas.

3. The highest and lowest relative peak intensities (peak areas) were observed for JA-Ile/Leu (average of 437.6 at 100 μM MeJA) and for JA-Phe (average of 6.6 at 100 μM MeJA). Similarly to this observation, JA-Ile was found to be the most prominent in *Arabidopsis thaliana*, with only minor amounts of JA-Val and JA-Phe present [21, 24].
4. JA is converted into JA-amino acids (JA-Ile, JA-Val and JA-Phe) and JA-hexose. The relatively high conversion to JA-Ile re-emphasizes its important role as a signaling molecule in plant adaptation. Other conversions might occur in order to perform specific biological functions, be cleared from the cells, or for storage.
5. In addition, mass spectral data indicate the presence of hydroxylated and carboxylated inactive forms of JA (12OH-JA-Ile, 12OH-JA-Val, 12OH-JA-Phe, 12COOH-JA-Ile) as well as conjugates of JA and dihydroJA.

4 Notes

1. Use suitable high purity organic solvents for UHPLC and for high definition mass spectrometry.
2. For elicitation studies with the aim of manipulating secondary metabolite pathways, non-physiologically high concentrations of MeJA of 100 μM and higher are often used [13].
3. It is essential to perform the analyses involving three biological replicates and three technical replicates ($n = 9$) to compensate for biological variability and instrumentation responses.
4. Methanol is a suitable organic solvent for the extraction of mid-polar and some polar metabolites.
5. Low temperatures and methanol as extraction solvent will quench metabolic activity, essential to eliminate possible enzymatic reactions occurring after harvesting and upon cell lysis.
6. All sample preparation steps should include a 0.2 μm filtration procedure to remove possible particulate material that can lead to blocking of the chromatographic columns.
7. The data from the low collision energy (CE) MS acquisition typically offer parent ion information (recquired for calculation of molecular formulae), while the data from the high CE MS scans offer fragment ion information (useful for structural elucidation).
8. Base Peak Intensity (BPI) chromatograms represent the intensity of the most intense peak at every point in the analysis. BPI peak chromatograms are cleaner chromatograms and more informative than total ion current (TIC) chromatograms,

because the background is reduced by focusing on a specific/single ion at every point.

9. Selected reaction monitoring (SRM) or multiple reaction monitoring (MRM) utilizes tandem MS (MS/MS) where a specific product ion from specific parent ion is selected/analyzed. The parent analyte is first selected, while other ions are filtered away. The parent analyte ion is then fragmented in the gas phase and a specific fragment ion is (transition) is monitored. This experiment has very high specificity because the SRM chromatogram represents only ions of a particular mass that fragment in a manner that produces a very specific product mass.
10. If authentic standards are not available, the Metabolomics Standards Initiative (MSI) [25] recommends the use of the terms 'annotation' or 'putative identification', indicated as MSI level 2.
11. MeJA and JA exist as four stereoisomers that result from the two chiral centers (at C-3 and C-7) present in the chemical structures. These isomers are (+)-(3R,7S), (−)-(3S,7R), (+)-(3S,7S), and (−)-(3R,7R) forms [26, 27]. The naturally occurring plant JA stereoisomers have been reported to be the (−)-(3R,7R) and (+)-(3R,7S) forms that can interconverted through epimerization [26].

Acknowledgments

The University of Johannesburg and the South African National Research Foundation are acknowledged for fellowship support to CHD and grant support to IAD [grant number 95818]. Dr. PA Steenkamp is thanked for UHPLC–MS analyses, Ms. N Ndolvu for assistance with LC/MS/MS analyses, and Dr. F Tugizimana for assistance with data analysis.

References

1. Okada K, Abe H, Arimura G (2015) Jasmonates induce both defence responses and communication in mono- and dicotyledonous plants. Plant Cell Physiol 56:16–27
2. Yan Y, Borrego E, Kolomiets MV (2013) Jasmonate biosynthesis, perception and function in plant development and stress responses. In: Lipid metabolism, vol 14. InTechOpen, London, pp 456–466. https://doi.org/10.5772/52675
3. Schaller A, Stintzi A (2008) Jasmonate biosynthesis and signaling for induced plant defence against herbivory. In: Induced plant resistance to herbivory. Springer, Dordrecht, pp 349–366
4. Wasternack C, Strnad M (2016) Jasmonate signaling in plant stress responses and development – active and inactive compounds. Nat Biotechnol 33:604–613
5. Glauser G, Grata E, Dubugnon L, Rudaz S, Farmer EE, Wolfender J (2008) Spatial and temporal dynamics of jasmonate synthesis and accumulation in Arabidopsis in response to wounding. J Biol Chem 283:16400–16407
6. Caarls L, Elberse J, Awwanah M, Ludwig NR, Vries M, Zeilmaker T, Van Wees SC, Schuurink RC, Van den Ackerveken G (2017) Arabidopsis jasmonate-induced oxygenases down-regulate plant immunity by hydroxylation and

inactivation of the hormone jasmonic acid. Proc Natl Acad Sci 114:6399–6393
7. Song S, Qi T, Wasternack C, Xie D (2014) Jasmonate signaling and crosstalk with gibberellin and ethylene. Curr Opin Plant Biol 21:112–119
8. Widemann E, Grausem B, Renault H, Pineau E, Heinrich C, Lugan R, Ullmann P, Miesch L, Aubert Y, Miesch M, Heitz T (2015) Sequential oxidation of jasmonoyl-phenylalanine and jasmonoyl-isoleucine by multiple cytochrome P450 of the CYP94 family through newly identified aldehyde intermediates. Phytochemistry 117:388–399
9. Woldemariam MG, Onkokesung N, Baldwin IT, Galis I (2012) Jasmonoyl-L-isoleucine hydrolase 1 (JIH1) regulates jasmonoyl-L-isoleucine levels and attenuates plant defences against herbivores. Plant J 1:758–767
10. Koo AJ (2018) Metabolism of the plant hormone jasmonate: a sentinel for tissue damage and master regulator of stress response. Phytochem Rev 17. https://doi.org/10.1007/s11101-017-9510-8
11. Zhang T, Poudel AN, Jewell JB, Kitaoka N, Staswick P, Matsuura H, Koo AJ (2016) Hormone crosstalk in wound stress response: wound inducible amidohydrolases can simultaneously regulate jasmonate and auxin homeostasis in Arabidopsis thaliana. J Exp Bot 67:2107–2120
12. Cimini S, Ronci MB, Barizza E, de Pinto MC, Locato V, Schiavo FL, De Gara L (2018) Plant cell cultures as model systems to study programmed cell death. In: Plant programmed cell death. Humana Press, New York, NY, pp 173–186
13. James JT, Tugizimana F, Steenkamp PA, Dubery IA (2013) Metabolomic analysis of methyl jasmonate-induced triterpenoid production in the medicinal herb, Centella asiatica (L.) Urban. Molecules 18:4267–4281
14. Mhlongo MI, Steenkamp PA, Piater LA, Madala NE, Dubery IA (2016) Profiling of altered metabolomic states in Nicotiana tabacum cells induced by priming agents. Front Plant Sci 7:1527
15. Hamany Djande CY, Piater LA, Steenkamp PA, Madala NE, Dubery IA (2018) Differential extraction of phytochemicals from the multipurpose tree, Moringa oleifera, using green extraction solvents. S Afr J Bot 115:81–89
16. Liu X, Yang Y, Lin W, Tong J, Huang Z, Xiao LT (2010) Determination of both jasmonic acid and methyl jasmonate in plant samples by liquid chromatography tandem mass spectrometry. Chinese Sci Bul 55:2231–2235
17. Balcke GU, Handrick Bergau N, Fichtner N, Henning A et al (2012) An UPLC-MS/MS method for highly sensitive high-throughput analysis of phytohormones in plant tissues. Plant Methods 8:47
18. Riet KB, Ndlovu N, Piater LA, Dubery IA (2016) Simultaneous analysis of defense-related phytohormones in Arabidopsis thaliana responding to fungal infection. Applic Plant Sci 4:1600013
19. Murashige T, Skoog F (1962) A revised medium for rapid growth and bio assays with tobacco tissue cultures. Physiol Plant 15:473–497
20. Wasternack C, Feussner I (2018) The oxylipin pathways: biochemistry and function. Annu Rev Plant Biol 69:363–386
21. Göbel C, Feussner I (2009) Phytochemistry methods for the analysis of oxylipins in plants. Phytochemistry 70:1485–1503
22. Eng F, Haroth S, Feussner K, Meldau D, Rekhter D, Ischebeck T, Brodhun F, Feussner I (2016) Optimized jasmonic acid production by Lasiodiplodia theobromae reveals formation of valuable plant secondary metabolites. PlosOne 11. https://doi.org/10.1371/journal.pone.0167627
23. Guranowski A, Miersch O, Staswick PE, Suza W, Wasternack C (2007) Substrate specificity and products of side-reactions catalyzed by jasmonate:amino acid synthetase (JAR1). FEBS Lett 581:815–820
24. Staswick PE, Tiryaki I (2004) The oxylipin signal jasmonic acid is activated by an enzyme that conjugates it to isoleucine in Arabidopsis. Plant Cell 16:2117–2127
25. Sumner LW, Amberg A, Barrett D, Beale MH, Beger R et al (2007) Proposed minimum reporting standards for chemical analysis: Chemical Analysis Working Group (CAWG) Metabolomics Standards Initiative (MSI). Metabolomics 3:211–221
26. Han Y, Bai Y, Xiao Y, Du F, Liang Y, Tan Z, Zhao M, Liu H (2011) Simultaneous discrimination of jasmonic acid stereoisomers by CE-QTOF-MS employing the partial filling technique. Electrophoresis 32:2693–2699
27. Matencio A, Bermejo-Gimeno MJ, Garcia-Carmona F, Lopez-Nicolas JM (2017) Separating and identifying the four stereoisomers of methy jasmonate by RP-HPLC and using cyclodextrins in a novel way. Phytochem Anal 28:151–158

Index

Antony Champion and Laurent Laplaze (eds.), *Jasmonate in Plant Biology: Methods and Protocols*, Methods in Molecular Biology, vol. 2085, https://doi.org/10.1007/978-1-0716-0142-6,

Printed by Printforce, the Netherlands